OXFORD MEDICAL PUBLICATIONS

The Management of Chronic Pain

The Management of Chronic Pain

A. W. Diamond

and

S. W. Coniam

Consultants in Anaesthesia and Pain Relief,
The Pain Clinic,
Frenchay Hospital,
Bristol

Oxford New York Tokyo
OXFORD UNIVERSITY PRESS

Oxford University Press, Walton Street, Oxford OX2 6DP
Oxford New York Toronto
Delhi Bombay Calcutta Madras Karachi
Kuala Lumpur Singapore Hong Kong Tokyo
Nairobi Dar es Salaam Cape Town
Melbourne Auckland Madrid
and associated companies in
Berlin Ibadan

Oxford is a trade mark of Oxford University Press

Published in the United States
by Oxford University Press Inc., New York

First published 1991
Reprinted 1992, 1993

A catalogue record for this book is available from the British Library

Library of Congress Cataloging in Publication Data
Diamond, A. W. (Andrew William)
The clinical management of chronic pain/A. W. Diamond and
S. W. Coniam.
(Oxford medical publications)
Includes index.
1. Intractable pain. 2. Pain clinics. I. Coniam, S. W.
II. Title. III. Series.
[DNLM: 1. Pain—therapy. 2. Palliative Treatment. WL 704 D537c]
RB127.D55 1991 616'.0472—dc20 91-3011
ISBN 0-19-263002-4

Printed in Great Britain by
Bookcraft Ltd.
Midsomer Norton, Bath

To our wives, Sally and Patsy, our children, our staff, and our patients

For their tolerance

Preface

This book is about the management of chronic pain. It is not a comprehensive description of the diagnosis and relief of all pains; that would require a comprehensive textbook of all medicine. Conventional medical practice leaves a lot of pain unrelieved. This pain becomes chronic and its management has become a new specialty of medicine. This field of medicine is drawing to it physiotherapists, occupational therapists, nurse specialists, and clinical psychologists. The doctors managing the clinics are largely anaesthetists. Because the concept of a pain clinic is so new, training is not yet established. The pattern of work is, however, becoming clear. It is now possible to describe in a textbook how to handle the clinical problems that will confront the workers in a Pain Clinic and the conditions that a Pain Clinic rather than some other discipline of medicine should deal with.

This textbook is about Pain Clinic work and we hope it will prove helpful to those who are starting to work in a clinic as consultants, to those training in pain management, and to those whose curiosity is driving them to ask precisely what a Pain Clinic does.

Bristol   A. W. D.
November 1990   S. W. C.

Acknowledgement

Line illustrations by Rosemary Moreton R.G.N., formerly Staff Nurse to the Frenchay Hospital Pain Clinic.

Contents

1 Pain Clinics and pain relief

Introduction

If you set up or work in a Pain Clinic you are, in effect, making a statement to the patient suffering from pain that 'We, the clinic, have the means to relieve pain and will provide pain relief to you, the sufferer.' Whatever the realities of the situation this is what your patient will believe; and quite reasonably, since after all this is what the rest of medicine is saying about its role in sickness and health. Whatever the future may reveal, at present we know that this expectation is hugely optimistic. We can relieve some sufferers of their unrelieved intractable pain; others can have their suffering reduced; but for many their pain will be unaffected even though we bring all the currently available skills and knowledge to bear on their condition. This failure to fulfil expectations imposes some of the strongest pressures on Pain Clinic workers that are experienced anywhere in medicine, yet if we succumb to it by giving up because we deem the situation hopeless, what good have we done anyone? By the same token, if we are driven by the patient's demand for pain relief further and further along the path of risk, for an increasingly diminishing hope of benefit until eventually we cause an increase in damage, pain, and disability, we have failed in the worst way possible. The attitude that we must have and that we must communicate to our patients and the world at large is that a Pain Clinic is not invariably a Pain Relief Clinic but often a Pain Management Clinic in which every effort will be made, not only to relieve the feeling of pain, but also to reduce the suffering and disability that it causes.

We must know as much about pain perception as we can. We need to analyse our patient's problem in the light of such knowledge. We must apply only those pain relief systems that seem to have some foundation in scientific method and do it systematically and we must bear in mind the balance that must be maintained in chronic pain sufferers between risk and benefit, especially where the only symptom is pain and there is no physical evidence of organic disease. To get as near to this ideal as we can, we must be systematic in our approach. This book will attempt to describe such a way of dealing with pain in clinical practice, first of all by looking at the problem of pain perception and then by relating it to intractable pain as it presents clinically and to current methods of pain relief and management.

We should first consider if the relief of pain as an end in itself is a right one.

Good, Western, scientific medicine, the only system of medicine that has brought consistent advances in the cure of disease, teaches that symptom relief comes from an understanding of the cause of symptoms and the relief of symptoms by eradication of that cause. To treat a perforated peptic ulcer, an inflamed appendix, or meningitis solely by the relief of the pain that they cause would be disastrous. Good medicine requires a systematic analysis of the patient's symptoms in the light of their past medical history, a physical examination, appropriate investigation, then the use of the sciences of pathology and therapeutics to eradicate the cause of the patient's disease, resulting in the relief of symptoms. This is the gold standard of medicine. Any patient, with any complaint, must be subjected to this analytical process and where pathology can be identified, where a disease can be eliminated, this must be the primary aim. Where there is an uncertainty in the pain clinician's mind about the presence or absence of treatable disease, then before pain relief becomes the only target, a diagnosis must be established (or be found impossible to establish) by the most informed source available. Multidisciplinary has many meanings in the pain relief world; an essential one is the recruitment of any medical discipline to eliminate the inappropriately early introduction of symptom relief as an end in itself.

Accepting this, modern medicine fails to provide us with a logical understanding, and from it logical therapy, for a vast proportion of the complaints for which sufferers request relief. Migraine and other headaches, the arthritises, dysmenorrhea, backache, malaise, muscular pain, irritable bowel syndrome, Crohn's disease, and ulcerative colitis, and the degenerative diseases of the central nervous system all defy a pathological basis. Then even where logical pathology gives us this understanding and, in theory, the potential to 'cure', of course delivery of the 'cure' can by no means be guaranteed. Untreatable cancer comes to mind first, but the symptoms of diabetes, vascular insufficiency, postherpetic neuralgia, and thalamic pain will often not be relieved by any possible treatment, where it is possible, of the disease no matter how clearly the pathology is understood.

Of all the persistent symptoms of disease, pain is the most demanding. It signifies damage and thus cannot be neglected; the sufferer's instinct induces pain actions and behaviours which affect not only their own life but also the life of those around them.

Our attitude to pain has been influenced by our history and culture. Before modern surgery had begun to develop, following the discovery of anaesthesia, disease was imposed by fate and unless trivial often led to death. The relationship between therapy, recovery, and even feeling better was tenuous. The development of surgery led to the concept that illness was a sign of something wrong inside and that this could be put right.

Alcohol and opium have been available from time immemorial. The place of alcohol sedation in distress in Western cultures is clear from novels and the treatments of primitive medicine. Opium was more a drug to induce sleep and subsequently to induce dreams. This relationship between opium and pleasure led to the idea of the dangers of opium which is still widely held and believed. Used by poets and writers such as de Quincey and Coleridge at the end of the eighteenth century, the Victorians regarded it with puritanical disfavour, preaching that it destroyed moral spirit. By the end of the nineteenth century it was known to be highly undesirable, its use almost certain to result in addiction and moral decline. The search for morphine alternatives led to no improvement in the relief of pain. Cocaine was described as an agent that would produce surface anaesthesia by Freud, who was looking for a substance whose stimulant properties would be an antidote for morphine addiction! Diamorphine was introduced as a non-addictive cough suppressant substitute for morphine. There are still those who believe that pethidine is less addictive than morphine and the most recent textbooks on the subject advocate the use of the less powerful opioids, codeine and dextropropoxyphene, in the relief of chronic pain where possible on the entirely specious grounds that in equipotent doses they will cause fewer 'problems' (by which is meant addiction) than morphine.

The search for analgesics outside the opiate group started with acetyl-salycilic acid (aspirin), originally thought of only as an antipyretic alternative to the expensive and toxic quinine. Its value as an analgesic has been difficult to assess as the vast majority of aspirin consumed is bought and taken without medical supervision. Further confusion has been caused by the introduction of the concept, which has a lot to do with encouraging sales of aspirin and its relations in arthritis, that drugs in this group are primarily of value because of an anti-inflammatory effect and that it is this that relieves pain. The huge demand for arthritis remedies has given rise to extensive research into the anti-inflammatory properties of the aspirin-like drugs, grouped significantly as the non-steroidal anti-inflammatory drugs. The concept of them as primarily analgesics seems to have attracted less attention, even though the safest drug in the group, with an analgesic and antipyretic potency equal to the others, paracetamol, has virtually no anti-inflammatory activity.

The search to avoid the use of morphine in severe pain led soon after the discovery of local anaesthetics in the 1880s to attempts to use them to block the nerves conducting chronic pain. It was but a small step to attempt to extend the techniques being developed in local anaesthesia to nerve destruction by neurolytic agents such as alcohol. Until the late 1940s there were very few anaesthetists, so that they were extremely busy and few had acquired the skills necessary to attempt neurolytic nerve blocks. By the middle of the 1950s the specialty of anaesthesia was growing in Great Britain and the United States. Maher in England introduced a more practical method of nerve ablation using phenol in 1955. Bonica in the United States developed, within a University Department of Anaesthetics, a Pain Clinic. From these beginnings a nucleus of clinicians started to take an increasing interest in Pain Relief as an end in itself

Physiologists had investigated pain perception from the days of Descartes but had been hampered by two barriers, the concept of pain as an injury detection mechanism alone and the consequential quest for a peripheral detection mechanism and an ascending pathway to a cerebral pain centre. Debate was concentrated around whether pain is perceived from a pattern of incoming stimulus detected by all types of sensory receptor and signalled to the brain when it has exceeded some threshold of intensity or through the stimulation of specific pain receptors, nociceptors. Of course, these concepts did not take into account the fact that pain can occur in the absence of tissue damage, that the experience of pain from any injury will depend upon many factors as well as the extent of the injury, and that there are many factors involved in the perception of pain beyond the main sensation, such as the motor and emotional responses to it. The work of Noordenbos (1959, 1964) and others culminated in the concept of a gate theory of pain perception described by Melzack and Wall in 1965. The theory sought to explain the fact that the brain can, through inhibitory or facilitatory influences on the dorsal horn of the spinal cord, 'tune' perception of pain so that a stimulus may at one time be painful and at another be perceived as painless. It also described the perception of pain as being a function of a pain 'action centre' involving all parts of the conscious brain, including emotional and motor responses. Melzack and Wall showed that peripheral tissue damage fires specific receptors, the nociceptors, and impulses are conducted through the smallest sensory nerve fibres to their cell bodies in the dorsal root ganglia and on into the dorsal horn of grey matter of the spinal cord. It is at this point that influences coming from other parts of the sensory nervous system, and descending control from the brain, influence whether this 'tissue damage' information will be signalled on up to the brain as pain. This explained the way in which the perception of pain is entirely psychological, and that the reception of pain impulses by the conscious mind is influenced by the activity of the brain and sensory impulses of other types. It generated an understanding of pain perception which has influenced pain management and research ever since. It introduced and formalized the introduction of the various stimulation pain control techniques, it helped the introduction of acupuncture as a pain relief method, and it explained the mechanisms by which the encephalins, endorphins, and the large number of other neuropeptides might work.

As Pain Clinics became more common in the United Kingdom and Ireland, The Intractable Pain Society was formed in 1966 initially as a forum in which clinicians could discuss and share the scarce information that was available; it subsequently attempted to formalize the training and facilities available to those working in pain relief. In 1972 the International Association for the Study of Pain was formed. This body, with its national chapters, seeks to encompass all those involved in the management of, and research into, pain. It produces the journal *Pain* and, through its chapters, has done a vast amount since its foundation to unify ideas throughout the world on pain research and to encourage the formation of Pain Clinics and the relief of pain in those areas in which these topics do not get the attention they deserve. It has also helped to unify the disciplines of medicine, psychology, nursing, and physiotherapy that are involved in the clinical management of pain.

In this textbook we seek to explain as far as possible the pain mechanisms involved in the conditions encountered in clinical pain relief practice and to follow a logical approach to their management.

References

Maher, R. M. (1955). Relief of pain in incurable cancer. *Lancet*, **1**, 836.

Melzack, R., and Wall, P. D. (1965). Pain mechanisms; a new theory. *Science*, **150**, 971.

Noordenbos, W. (1959). *Pain; problems pertaining to the transmission of nerve impulses which give rise to pain. Preliminary statement*, pp. 110–21. Elsevier, Amsterdam.

Noordenbos, W. (1964). Some aspects of Anatomy and Physiology of Pain. In *Pain* (ed. R. S. Knighton and P. R. Dumke), p. 249. Henry Ford Hospital International Symposium, Detroit, Oct. 21–23 1964. Little Brown and Co., Boston.

2 Pain perception

Pain is perceived in the cerebral cortex. This is the end of the journey that starts at a peripheral nociceptor, passes along an axon in a peripheral nerve, through the neuron of that axon, and onwards into the spinal cord. It continues up the cord through the hind- and mid-brain to the thalamus, and through the thalamo–cortical connections the analysed and processed information reaches consciousness and perception. The signal that is interpreted as pain can originate at any point on this journey.

It is now accepted that a stimulus which is sufficiently intense to cause potential tissue damage fires specific receptors, the nociceptors. Earlier theories suggested that a sensation was perceived as pain when a particular pattern of stimuli reached the central nervous system (CNS) through non-specific receptors and pathways: this was analysed by the CNS with regard to intensity, space, and time and was interpreted as pain if its nature might cause tissue damage. However, the evidence that the impulses from damaging stimuli travel centrally along specific axons which do not conduct impulses from non-damaging stimuli is now overwhelming, thus indicating that the terminals of these axons must be specific detectors of damaging stimuli.

Stimulation experiments show that pain is conducted along two types of nerve fibre, one relatively fast-conducting and the other slow. The faster of the two is the smallest of the myelinated fibres—the A-delta, between 2–5 μm in diameter and conducting at a rate of 6–30 ms^{-1}. These fibres terminate peripherally in the receptors specifically for the detection of mechanical deformation, the high threshold mechanoceptors (HTM). Each fibre serves up to twenty receptors at discrete points over a small area of skin. The slower is the smallest fibre of all, the C-fibre, less than 2 μm in diameter and conducting impulses at a rate of 0.5–2 ms^{-1}. These fibres terminate peripherally in a single receptor which responds to all forms of noxious stimulation, the polymodal nociceptor (PMN). The skin is well studied and the place of these receptors understood because of their accessibility in human volunteers, but the nature and function of nociceptors in other tissues is more difficult to examine. Muscles and joints are supplied with HTMs, and PMNs are widespread in most other tissues. The relationship between fibre diameter and perception of noxious stimuli in viscera is not yet clear, investigation being hampered because the visceral innervation is sparse and its constant tonic firing is difficult to separate from that generated by noxious stimulation. The fibres arise from cells in the dorsal root ganglia 14–30 μm.

in diameter, the small B group. Some of the central axons of these neurons enter the cord through the anterior root but the majority enter in the ventrolateral bundle of the dorsal root. On entering the cord, many of the roots bifurcate, one branch ascending and the other descending the cord. These in turn give off collateral branches, most of which enter the grey matter of the dorsal horn in their segment of origin but many spread several segments up and down the cord.

At this stage it is worth emphasizing again the integrity of the whole nervous system, in particular with regard to pain sensation. The modulation of incoming impulses to control what is perceived in consciousness is a function of the whole sensory nervous system and not exclusively of a particular part such as the dorsal horn or the mid-brain. Of necessity, each area in the CNS in which the onward passage of impulses is modulated tends to be studied separately, but the interaction between the different parts of the sensory system is complete. It is essential that this is understood clinically, since attempts to relieve pain aimed at only one part of the system, such as peripheral nerve or the dorsal root entry zone, are likely to fail because of this interdependence.

Rexed (1952) divided the dorsal horn into laminae separated by their morphology. These laminae interconnect and interact to a considerable degree but do have separate functions that relate to pain perception and its control. Lamina I, the dorsal lamina, and to a lesser extent its ventral neighbour Lamina II, appear to be particularly related to the input from nociceptors in having many cells that respond to potentially damaging stimuli although even in these laminae the cells are not exclusively concerned with nociception. In Lamina III there are large numbers of fibres from the deeper cells. The cells in this lamina and those of Lamina IV become increasingly excitable by non-noxious, low threshold stimulation.

Lamina V cells send a large number of their efferent fibres to the brain. Some of them respond to visceral fibres and at the same time have a small area of cutaneous receptive field. As this would correspond with the way in which visceral pain is perceived it suggests that these nerves subserve visceral nociceptors. Other Lamina V cells have a small central area of their receptive field in which they are stimulated by low threshold stimulation and a larger peripheral receptive field in which they respond only to high threshold noxious stimulation. Yet other cells in the deeper laminae respond to both low and high threshold stimulation, the wide dynamic range neurons. These cells can fire in the presence of peripheral nerve damage to the release of noradrenalin, and their response is reduced by noxious stimulation elsewhere in the body (diffuse noxious inhibitory control, or DNIC). While the other cells in Laminae I, II, and V appear to be related to transmitting information about tissue damage, these cells relate to the perception of neurogenic pain.

The cells in Laminae I, II, and V interact with one another in a cascade fashion. Lamina II contains cells that are excitatory to cells in Lamina I, and Lamina II there are also cells that transfer C fibre input to Lamina V. All laminae except Lamina II send fibres to the brain, suggesting that the cascade system passes inwards so that each lamina receives an increasingly complex pattern of stimulation. From Lamina III inwards, the brain is sent information to analyse that is a summation of the information that cells in the lamina have collected from the laminae dorsal to it and also from direct input from sensory neurons whose axons go to the periphery. However, what they will transmit onwards into the nervous system depends not only on the excitation that they are receiving from outer laminae and the periphery, but also on inhibition arising in the dorsal horn and descending from higher up in the nervous system.

It is now beyond doubt that this excitation and inhibition is humorally mediated. The number of possible neurotransmitters, discovered using the techniques now available such as radioimmunoassay, suggests an extremely complex system. Much of the unravelling of the way in which the system may work has resulted from the discovery of the endorphin system. It had become clear that morphine must be the ligand for specific receptors in the CNS and Hughes *et al.* (1975) developed two models to detect this activity, the guinea pig ileum and the rat vas deferens. Using material extracted from pig brain they were able to identify two pentapeptides produced within the CNS which had opioid activity; they named these encephalins. One of them contained leucine and stimulated a morphine receptor in the ileum model, the mu receptor, and was named *leu* encephalin. The other stimulated the rat vas deferens, the delta receptor, and was named *met* encephalin as it contained methionine.

Since then, two other opioid receptors have been identified, the kappa, whose identification followed the development of ketacyclazocine, and the sigma, the ligand for which is SKF 10,047. The identification of these separate receptors has not, in spite of the most vigorous efforts, yet produced a clinically useful application. However, encephalins which are ligands for these receptors are found in the dorsal horn and form part of the endogenous inhibitory system. Substance P, another peptide which is plentiful in the dorsal root ganglion neurons of C fibres and in the parts of the dorsal horn in which their afferent fibres terminate, may be the chemical transmitter of tissue damage information. As well as the endogenous opioids, within the dorsal horn, noradrenalin and serotonin are inhibitory to nociception and to other peptides such as cholecystokinin; vaso active intestinal peptide (VIP), and somatostatin affect nociceptive transmission, and possibly have a function related to the perception of visceral pain.

Inhibition of nociceptive transmission within the dorsal horn would appear to be the principal way in which we achieve analgesia to nociceptive pain in clinical practice (Basbaum and Fields 1984). In experimental animals such inhibition may be generated within the dorsal horn endogenously by peripheral stimulation. It may be induced by generating descending inhibition of pain perception in pathways that originate in the brain or elsewhere in the cord, or by pharmacological means using drugs that act on the receptors of the endogenous opioids.

Two types of peripheral stimulation can be applied segmentally: high intensity, low frequency stimulation of the acupuncture type and low intensity, high frequency stimulation, as generated by massage or transcutaneous electrical stimulation. A third type of stimulation-induced analgesia is diffuse noxious inhibitory control, in which a painful stimulus anywhere in the body induces a reduced ability to feel pain in other areas (Le Bars *et al.* 1979). There is evidence that these three stimulus-generated analgesia mechanisms are different from each other. High intensity, low frequency stimulation is reversible by the morphine antagonist, naloxone, whereas low intensity, high frequency stimulation is not. These mechanisms both have a segmental basis while diffuse noxious inhibition does not.

At dorsal horn level then it is probable that the analgesia induced by LFHT stimulation is the result of endogenous opioid production (He 1987), probably by cells within Lamina II. HFLT stimulation was introduced deliberately to invoke the peripheral stimulatory analgesia system and was thought to induce analgesia initially, as a result of presynaptic inhibition. Anatomical and pathological changes in the nervous system that should have fitted in with this theory have never materialized and it is now thought that the mechanism for this type of stimulus-induced analgesia must be humoral as well. It is currently thought that it may be due to a glutaminergic mechanism but evidence to confirm this is lacking.

Ascending fibres from the dorsal horn nociceptor neurons cross to the other side of the cord and ascend in the spinothalamic tract to the brain. This was thought to be the only way in which pain information reached perception in humans, and this belief was confirmed by the fact that surgical interference with the anterolateral quadrant of the cord or its interruption by other destructive means interfered with the ability to feel pain. Destruction of these fibres has been used for some time for pain control in those unlikely to survive for long. In the small number of patients where the technique has been used to provide pain relief in long-term survivors, an altered type of pain perception has developed, in some in a matter of weeks and in others in a matter of months. Discrimination is not so accurate in these subjects, and the interval between a painful stimulus and its perception is longer. This suggests that in the absence of the rapid, highly discriminative pathway, probably of recent development, a slower, older pathway will begin to function.

The cells from which the spinothalamic tracts originate send collaterals to several other tracts which are known to be involved in pain perception. The spinoreticular pathway is involved in the motivational and affective aspects of pain and also the autonomic response; the spinomesencephalic tract may share this function, and its destination in the periaqueductal grey matter is adjacent to the cells which generate descending inhibition of pain perception. The return of pain perception following a lesion of the spinothalamic tract may be through ascending spinocervical and second-order dorsal column pathways, though little is known of the destination of fibres running in these tracts. The spinothalamic tracts terminate in the ventral posterior lateral nucleus and the central lateral nucleus of the intralaminar complex. From here information is sent to the sensory cortex. Fibres ascending in the spino-reticular pathway terminate in the lateral reticular nucleus, a cerebellar relay nucleus, and also in the nucleus gigantocellularis and the lateral pontine tegmentum. Fibres ascending in the spinomesencephalic pathway terminate in the superior colliculus and the intercollicular nucleus, as well as in the periaqueductal grey matter and adjacent reticular formation.

In summary, the spinothalamic tract communicates with the thalamus which relays its information to the cortex, through which pain is perceived in the intact consciousness in all its detail. The other pathways whose destination is understood would appear to have some function with regard to the generation or inhibition of impulses in the controlling inhibitory pathways that originate in the mid-brain and terminate in the dorsal horn.

Pain perception occurs through an 'action system'. The immediate response is that of warning and protection. This is followed by withdrawal, vocalization, an autonomic response, and examination of the damaged area. Analysis of the perceived event in the light of memory then occurs followed by the use of rubbing and other stimulatory actions to lessen the perceived pain. If this response is analysed it will be seen that many parts of the central nervous system are involved in this system, the motor cortex for the response of movement, the sensory cortex to permit discrimination of the nature of the stimulus, the limbic system and the frontal lobes as the source of the emotional response, the hypothalamus for the autonomic response, and so on. There is thus no 'pain centre' in the brain but rather the perception and response to pain are a function of the whole CNS.

This chapter attempts to explain the complexity of tissue damage detection and pain perception. In attempting to relieve any pain, when a lesion outside the nervous system is the obvious source or when there is no evidence of tissue damage, parts or all of this complex and interacting system will be functioning. It is because of this complexity that attempts to relieve pain, even when its cause is obvious, can fail. Where the pain has become chronic, where the CNS itself has become damaged and this damage is generating pain, or where pain's presence has generated behavioural changes, the relief of pain becomes part of an immensely complex problem and searching for some simple localized solution peripherally or centrally is doomed to disappointment.

References

Basbaum, A. I. and Fields, H. L. (1984). Endogenous pain control systems: brain stem spinal pathways and endorphin circuitry. *Annual Reviews in Neuroscience*, **7**, 309–38.

He, L. (1987). Involvement of endogenous opioids in acupuncture analgesia. *Pain*, **31**, 99–122.

Hughes, J., Smith, T. W., Kosterlitz, H. W., Fothergill, L. A., Morgan, B. A., and Morris, H. R. (1975). Identification of two related pentapeptides from the brain with potent opiate agonist activity. *Nature*, **258**, 577–9.

Le Bars, D., Dickenson, A. H., and Besson, J. M. (1979). Diffuse noxious inhibitory controls (DNIC). I. Effects on dorsal horn convergent neurones of the rat. *Pain*, **6**, 283–304.

Maher, R. M. (1955). Relief of pain in incurable cancer. *Lancet*, **1**, 836.

Rexed, B. (1952). The cytoarchitectonic organisation of the spinal cord in the cat. *Journal of Comparative Neurology*, **96**, 415–95.

3 Assessment of the pain patient

Patients are referred to the Pain Clinic by hospital colleagues, general practitioners, and occasionally other hospital staff who wish to have assistance in managing patients in pain. However, colleagues may have widely varying opinions about what they expect the Pain Clinic to achieve, and expectations which they in turn pass on to the patient. It may be that the patient is suffering from a well-recognized pain syndrome, such as causalgia or post-herpetic neuralgia which the colleague knows to be treated by the Pain Clinic, or sometimes the referral is a desperate attempt to get help for the distressed patient for whom all pain-relieving drugs known to the colleague have been found to be ineffective. Unfortunately, the referral may sometimes be a means of disposing of problem patients who will not respond as expected by another practitioner, so that other outpatient departments can be left free to see patients with 'real' diseases. The assessment of the pain patient must, therefore, be made with reference to previous treatments and the way in which symptoms have been explained to the patient (if at all) by other practitioners. The patient who has seen a non-medical alternative practitioner previously, and been given a neat, but perhaps unfounded, account for symptoms, must be assessed differently from the patient who has been impatiently dismissed elsewhere with 'It's all in your mind, dear', because a 'medical' cause for symptoms cannot be discovered.

Pain clinicians have different ideas of their own roles. At one extreme, some will see any patient who complains of pain as a presenting symptom, and consider that their medical training and experience are adequate to investigate all aspects of the problem. At the other extreme, some only see themselves as nerve block technicians who perform tasks requested by other doctors, or who will only work as part of a multidisciplinary team. The ideal is probably somewhere in between. Certainly the pain specialist should feel competent to assess the overall medical condition of the patient, and to diagnose pathological causes of pain. However, the pain specialist has usually received the majority of his or her training in another field, and should recognize when it is appropriate to enlist the help of colleagues experienced in other branches of medicine, in order to further investigate or treat specific problems. It remains a point of debate as to whether a Pain Clinic should see patients as a primary referral in all cases or whether, in some instances where pathology has not been previously demonstrated, patients should only be referred after investigation by a specialist in the system in which the symptoms are apparent. The latter attitude is taken to an extreme by some clinics where GP referrals are not seen at all. The pain practitioner must form his or her own opinions on this matter, in the light of training and experience, as well as the nature of relationships with colleagues in other specialties.

Whatever the mode of referral to the Pain Clinic, it is essential to identify symptoms and signs which do require further investigation, before treating pain as an isolated symptom. Inadequately investigated pathology should be further elucidated before treatment as, although it should not be exclusive to the management of pain, the risk of allowing treatable pathology to progress while symptoms are managed would be unacceptable. It is only if the pain specialist is satisfied that all possible pathological causes have been adequately considered that management on a purely symptomatic basis can continue. Conversely, a Pain Clinic is not fulfilling its true role if it will only treat when a pathological diagnosis can be made. The contribution of the specialty of pain relief to patient care has been to recognize that much chronic pain defies pathological diagnosis in classical disease terms, with the present state of knowledge of the pathophysiology of pain, but this does not remove the need to attempt to provide adequate pain management. Pain is not a measurable, definable entity, but a (usually) unpleasant experience which is personal to the sufferer. It is, therefore, what the patient says it is, and true fabrication of pain, or malingering, is probably not common. We may not feel that a patient's symptoms fit with our own impression of their physical health, or we may feel that they are reacting excessively to what we feel can only be a minor symptom. Nevertheless, if the patient is distressed by an experience which he or she describes as pain, then that distress, or its cause, is the reason for them seeking help, even if it is beyond our medical ability to understand or relieve that distress. We may even feel that the patient's symptoms are a function of their personality or a reaction to their environment, but to that patient, the experience of pain is as real as from any other cause. There may be an indication to attempt to adapt the way in which the patient views symptoms of pain, or their cognitive processes, rather than applying yet more traditional medical models of treatment.

A patient's disability can be considered to be a sum of the physical component plus the patient's emotional reaction, with the addition of environmental, social, or employment factors. Any one of these may predominate, and the patient whose symptoms are felt to result from a minor physical component (which may be primary or secondary) plus a large abnormal emotional response, should perhaps have treatment directed at the latter before excessive physical treatment or investigations are attempted.

Some patients require a more expert psychiatric or psychological assessment. The pain specialist should be able to decide which patients to refer for specialist advice, even if it is beyond his or her expertise to perform that assessment fully. However, even patients with pronounced psychiatric symptoms may still require help in managing their pain symptoms.

Many patients with chronic pain have spent years in seeking the 'Holy Grail' of a cure for their symptoms. Doctors or alternative practitioners previously seen may have told the patient that, although they have not been able to help, further investigations or specialist consultations should be sought. Although most practitioners feel that the patient should be given every possibility of finding a 'cure', some patients embark on this odyssey with ever more extensive (and negative) investigations and treatment. Perhaps surprisingly, this further reinforces the patient's belief that there is a yet to be discovered disease from which they are suffering, or even that the doctors are concealing the true nature of their condition. An important function of the Pain Clinic is sometimes to tell a patient that a pathological cause for their symptoms is not going to be found, and that the pain must be considered as a disorder in its own right. The journey around various specialist clinics should therefore cease, and energies should instead be directed to coping with the pain and the effects it has on the patient's life, and returning responsibility to the patient. Ending the passive role of the pain-suffering, doctor-dependent patient is a major step in helping to relieve the patient's suffering.

An important part of the Pain Clinic interview is to assess the patient's understanding of the symptoms and expectations of treatment. A patient should be helped to reach an understanding of the nature of chronic pain, and what are the realistic expectations of relief. Many patients with chronic pain are angry and frustrated at the medical profession's inability to diagnose the cause of their pain. To give the pain syndrome a label, even if it has little scientific basis, often provides relief to patients and is helpful in enabling them to come to terms with their problems. However, it is important not to provide a diagnostic label which implies that theirs is a readily corrected defect. If the doctor believes that there is a large psychogenic component to the pain, presenting this view bluntly can antagonize the patient who sees this as the equivalent of being labelled as mad. This only reduces the patient's ability to cope with the pain, and encourages them to seek a physical 'cure'. Psychogenic aspects of pain must be approached gradually and in terms acceptable to the patient.

When assessing a patient with chronic pain, it is important to establish the following:

1. The broad nature of the complaint. How and when did it start?
2. What investigations have been done? Were there any relevant findings?
3. What treatments have been tried in the past, or are still being used? What effect have they had on the symptoms?
4. The nature of the pain, in all its aspects.

5. The effect of the pain, in all its aspects. Is there an apparent disparity between the patient's behaviour/emotional response to the pain, and the objective symptoms and signs?

The history

Find out if the pain followed any physical or psychological events or trauma in the patient's life. Is there a history of injury, surgery, or other illnesses? Has the pain been continuously present, or are there pain-free hours, days, or weeks? Are there any obvious factors which might account for these fluctuations? Is the pain at a constant level, getting worse or better with time, or are there diurnal variations? Has the pain been investigated previously; if so, how, and what were the findings? What has the patient been told previously about the pain? What operations, drugs, physical treatments, and so on have been applied, and what were the results of such treatments? It is also important to enquire about any other major items of the patient's medical, psychiatric, social, and employment history.

The pain

Enquiries about the nature of the pain have three main purposes. Firstly, the patient's description of symptoms may be helpful in diagnosing the cause of the pain. Secondly, a description of the type of pain may be important in choosing the most appropriate treatment for that symptom. Thirdly, it may give the clinician an idea of the patient's reaction to the pain, and the effects which it has on his or her life.

The patient's impression of the site of pain is asked, and the direction(s) of radiation, if any, noted. If multiple sites are described, it may be helpful if the patient draws a picture to illustrate where pain is felt, or to indicate this on a line drawing. If the patient is requested to use different types of shading or symbols on a body outline to describe different pains, the clinician may also learn a good deal of information about the patient's psychological reaction to the symptoms. Bizarre diagrams, often heavily annotated with emotional comments, and overloaded with site indications may suggest that some form of psychological assessment may be appropriate.

Enquiries should be made about aggravating or relieving factors, including movements and daily living activities. These may provide helpful clues to the best means of alleviating symptoms. It is also helpful to know the time course of the pain in relation to daily activities.

Particularly important is to request a description of the pain. This may have already been offered in elaborate terms, but frequently patients say that they cannot describe pain. Suggest a few terms, and the rest often flows spontaneously. It is useful to record verbatim some of the expressions used to describe the pain, as these can be valuable in determining the nature of the pain, and the response to that pain. This has been utilized particularly in the form of the McGill Pain Questionnaire.

The McGill Pain Questionnaire utilizes a list of words describing pain qualities. They involve three major groups representing sensory qualities (in terms of temporal, spatial, pressure, and thermal qualities), affective qualities of the experience (in terms of fear, tension, and so on), and evaluative terms that express the patient's subjective experience of the pain. These groups are divided into subgroups of similar properties, but perhaps differing in nature or intensity. The words were given numerical scores in the original question-naire by Melzack and Torgerson (1971), which expressed intensity as generally agreed by a wide range of the population of North America. The system has been widely modified and adapted, but it can be used as a basic tool for verbal and numerical assessment of pain quality and intensity (Melzack 1975).

The questionnaire may be useful for specific investigations and trials, but it is not always necessary in day-to-day patient assessment. It should be possible to detect a strongly emotive response to pain from the patient's description. Words such as 'punishing' or 'cruel' say more about the patient than the pain. It is especially important to determine whether the pain has a lancing/shooting/electric shock type quality, as this has specific implications for treatment. A description of 'raw', 'burning', or 'sore' pain may also help with the diagnosis and treatment. Does the patient complain of allodynia, hyperpathia, anaesthesia, or dysaesthesia? Frequently there may be several different components to the pain, and it is important to discover if this is the case, both from a diagnostic point of view, and in order to assess the effect of treatment.

The effects of the pain must be discussed to gain a true picture of the patient, to indicate whether other aspects of the patient or his environment need treating or assisting, and sometimes to help form a prognosis. Are there any physical or physiological effects of the pain, such as nausea, weakness, or insomnia? What is the effect of the pain on the patient's activity (very important for assessing response to treatment), social, domestic, or marital life, and employment and financial situations?

Enquiries should be made concerning the current treatment and medica-tion that the patient is receiving, and what the response to this has been. Frequently, pain patients have accumulated a vast amount of medication, with every practitioner consulted adding to the list. None have dared to risk removing what may be an essential prop. An important role of the Pain Clinic may be to rationalize drug therapy, removing medications which are obviously ineffective or duplicate the actions of other drugs. Some patients improve simply by reducing unnecessary medication along with its accom-panying side effects.

The physical examination of the patient with chronic pain is as important as with any other patient, even though there may be few, if any, physical signs to discover. The examination starts when the patient enters the room. Has this patient become withdrawn, depressed, and a total victim of a pain syndrome, or has he or she become aggressive, resentful, and mistrusting of the failed efforts of the medical profession to find a 'cure'? Frequently, a patient's coping ability becomes apparent at an early stage in the interview. Even if there are no physical signs, the examination, however brief, is important in developing the doctor–patient relationship. It helps to develop the patient's trust and confidence in the doctor, who takes the problem of pain seriously, and may even discover a treatable pathology. If there is no obvious pathology, then at least the doctor has been seen to confirm this.

The physical examination is often of value as a process of exclusion and may also help to confirm ideas suggested by the history, determine the extent of the problem, discover more about the patient's reaction to pain, and to look for provoked symptoms and signs, such as the discovery of tender trigger points, which may have been unsuspected by the patient.

Most of the physical examination of the patient with chronic pain follows standard medical techniques, and will not be described in detail here. If there appears to be a musculoskeletal problem, it is important to observe gait and other movements, as well as power, reflexes, and sensation. The patient is observed for abnormal function, be this a result of nervous or musculo-skeletal pathology, or conscious or unconscious limitation of movement as a result of pain. Is there an objective sign of nerve, muscle, or joint deficiency, or are the signs at variance with the symptoms?

A painful scar may present obvious changes of skin colour or texture, but any painful area of skin should be examined carefully for changes such as thinning, change of colour, or hair loss, which may indicate a sympathetic component to the pain. Tenderness is an important sign. This may be a superficial hypersensitivity or allodynia accompanying nerve damage, or may be deeper, arising from viscera, muscles, or connective tissue. If no local pathology is apparent in the area where the patient complains of pain, look for trigger points in the muscles which may refer pain to that area. Reference to works on myofascial pain (Baldry 1984; Travell and Simons 1983) will indicate muscles in which to search for trigger points. The muscles should be placed in a position of stretch, and palpated with the pad of the finger running over the muscle. A sudden response ('jump sign') by the patient may indicate a trigger point of which the patient was totally unaware, only having complained of the referred pain. The trigger point may be associated with a palpable nodule in the muscle, or a taut band along the direction of the muscle fibres.

Radiological investigations are often unhelpful in making a primary diagnosis of the cause of pain, unless there is some obvious pathological change in the skeleton, such as a fracture or metastasis. Even then, this may detract attention from other causes, and may be an incidental finding. However, radiology may be essential to exclude treatable pathology, or to confirm or localize disorders diagnosed from the history. It is a common pitfall to assume that a degenerative change seen on a radiograph is necessarily the cause of the patient's pain. Such changes are frequently present in patients who are symptom free, and absent in those with symptoms. Further assistance with diagnosis may include such procedures as computerized tomography (CT) scanning, thermography, and the use of diagnostic nerve blocks.

Another pitfall is that of narrowing one's view of the presenting symptoms to previously diagnosed conditions. This may at best lead to inability to treat the patient appropriately, or at worst, failure to see a serious and treatable alternative cause for the symptoms. For example, back pain may be referred from serious visceral, retroperitoneal, or aortic pathology, but attention may be diverted by a previous history of spinal degenerative disease, and/or marked radiological changes in the spine.

The full psychological assessment of patients should be performed by a clinical psychologist, although the pain physician often has to make a rough assessment of some aspects of the patient's psychological status. A clinical psychologist with an interest in chronic pain is an invaluable asset to a Pain Clinic. A number of questionnaires are available which can be scored by a non-psychologist to give some idea of psychological parameters. Widely used in Pain Clinics are personality assessment questionnaires (such as the Minnesota Multiphasic Personality Inventory (MMPI)), anxiety and depression questionnaires [Leeds (Snaith et al. 1976), Beck, The Hospital Anxiety and Depression Scale (Zigmond and Snaith 1983), and Zung], questionnaires which assess the impact of the illness on the patient (Follick et al. 1985), and assessments of abnormal pain behaviour. The McGill Pain Questionnaire has been modified in a number of ways, and it has been claimed that a high correlation can be found between those patients in whom symptoms are predominantly non-organic, and eventual negative findings, as predicted by the selection of words chosen (Leavitt and Garron 1979, 1980).

These are mainly of use in trials and research, to enable comparison of different patients, and their response to treatment. A clinical psychologist, or a well constructed questionnaire, may enable selection of patients who should be considered for treatment of anxiety and depression or other forms of psychiatric problem. They may also select patients who may benefit from psychological counselling, psychotherapy aimed at improving ability to cope, and behaviour or cognitive therapy as part of the management of pain.

The assessment of the amount of pain suffered by a particular patient is a problem which has always hindered pain research. Pain, being a subjective experience, is not measurable and measurements of applied painful stimuli in a laboratory are not always applicable to patients. What we can measure is the patient's subjective impression of the severity of the pain, in comparison with similar measurements after treatment. The usual means of doing this are by visual, numerical, or verbal analogues. The visual analogue scale (VAS) is a 10 cm line. At one end is 'no pain at all', and at the other end, 'the worst pain imaginable'. Patients are asked to mark the line at the point which they feel is representative of their pain at that time, and the record is then converted to a numerical score by measuring the distance of the mark along the line. Although this is perhaps one of the most frequently used measurements of pain, it is extremely crude, even if the patient fully understands this abstract concept (and this cannot be assumed!). There is often a tendency to mark the top end of the scale, even though this cannot be truly representative at the time of recording, and certain divisions of the line appear to attract patient's attention more than could be expected at random.

Patients can be asked to express pain intensity on a numerical scale from 0 to 10, but the same deficiencies occur as with the VAS, and certain numbers tend to be chosen more frequently. A verbal scale (such as: no pain, mild, discomforting, distressing, horrible, excruciating) can be given to the patient, and the word chosen can then be converted to a numerical equivalent. This may give a better impression of the patient's subjective response to the pain and is possibly more valuable when asking a patient to compare pain levels on different occasions. One of the major problems with any of these pain rating scales is that it is extremely difficult to remember the intensity of pain, so that even if pain has been reduced with treatment, the same number or point on the scale may still appear to represent the pain intensity to the patient.

Perhaps of more value is to assess the effect that the pain has on important aspects of the patient's life, and compare this on different occasions, by using a system such as the Pain Disability Index (Tait *et el.* 1990). A patient who has increased his walking distance, time out of bed, social activity, or even returned to work, has clearly responded in some beneficial way to treatment, even if subjectively he feels that his pain intensity has not altered. Batteries of questionnaires concerning daily activities, mood changes, and aspects of illness behaviour may be necessary to really demonstrate objectively the response to treatment.

Finally, it must be stressed that management of pain as a symptom should not lead to missing a curable disease. It is common to assume that the patient who has been extensively investigated with negative findings by many other doctors, and who complains incessantly of perhaps rather strange symptoms which do not readily fit with our experience, does not have a physical cause to account for the symptoms. This is often the case, but in this situation, it is essential to maintain a vigilance for any change in symptoms or signs which may suggest an underlying pathology, and require further investigation. The narrow path between missing curable disease, and encouraging the patient in his disease conviction and fruitless search for a cure, is a difficult path to tread.

References

Baldry, P. E. (1989). *Acupuncture, trigger points and musculoskeletal pain*. Churchill Livingstone, Edinburgh.

Follick, M. J., Smith, T. W., and Ahern, D. K. (1985). The sickness impact profile: a global measure of disability in chronic low back pain. *Pain*, **21**, 61–76.

Leavitt, F. and Garron, D. C. (1979). The detection of psychological disturbance in patients with low back pain. *Journal of Psychosomatic Research*, **23**, 149–54.

Leavitt, F. and Garron, D. C. (1980). Validity of a back pain classification scale for detecting psychological disturbance as measured by the MMPI. *Journal of Clinical Psychology*, **36**, 186–9.

Melzack, R. and Torgerson, W. S. (1971). On the language of pain. *Anaesthesiology*, **34**, 50–9.

Melzack, R. (1975). The McGill Pain Questionnaire. Major properties and scoring methods. *Pain*, **1**, 277–99.

Snaith, R. P., Bridge, G. W. K., and Hamilton, M. (1976). The Leeds scale for the self assessment of anxiety and depression. *British Journal of Psychiatry*, **128**, 156–65.

Tait, R. C., Chibnall, J. T., and Krause, S. (1990). The pain disability index: psychometric properties. *Pain*, **40**, 171–82.

Travell, J. and Simons, D. G. (1983). *Myofascial pain and dysfunction. The trigger point manual*. Williams and Wilkins, Baltimore.

Zigmond, A. S. and Snaith, R. P. (1983). The Hospital Anxiety And Depression Scale. *Acta Psychiatrica Scandinavica*, **67**, 361–70.

4 Back pain

Back pain is so common that it could be said to be part of normal human experience and no more a disease than dandruff. It is, however, pain: a signal of tissue damage is perceived and physical and psychological distress results. Pain in the back is common, in a local community of students 60 per cent had experienced acute low back pain in the preceding year. Eighty per cent of the population suffers from back pain at some time during life, and in the United States of America it is the third leading cause of disability, with 3.9 per cent of the population permanently disabled by it.

The problem is therefore huge and many in the medical and alternative areas seek to solve it. The first problem to overcome is that the terminology used by any one group suggests that the pathology is understood and fits in with the concept of the therapist, thus inferring that the diagnosed condition will be cured by the appropriate 'correct' therapy.

Pain from the low back frequently radiates to the legs or groin. It may be in the distribution of the sciatic nerve and called sciatica. The diagnosis of sciatica gives an impression of precision in this area, more so than lumbago or spondylosis (two other non-specific terms for radiating low back pain), but it is far from that. Does it refer to pain exclusively in the distribution of the sciatic nerve with no back pain element? How far down the leg does the distribution have to extend before the diagnosis of sciatica can be made: into the buttock, down the thigh to the knee, or must it go into the foot? If the diagnostic terms sciatica, lumbago, or spondylosis are used, the imprecision carries the great advantage that there is very little suggestion that the pathology of the cause of the pain is known. Other diagnostic labels carry an implication as to pathology.

The problem of attaching a pathological label that is difficult to sustain started with the advent of disc surgery in 1934 (Mixter and Barr, 1934). Once it was realized that intervertebral discs might herniate and that surgery to remove the herniated fragments might be followed by pain relief, it was natural that the rupture and pain should be considered cause and effect. The incidence of sciatica is so great that the operation has become immensely popular, as have manipulation and other techniques that claim to put the disc back in place. Some uncomfortable facts which do not seem to disturb the enthusiasts are: that herniated discs do not always cause back pain and sciatica, that there are no symptoms that are diagnostic of prolapsed intervertebral disc, that between 10 per cent and 50 per cent of patients will have symptoms that persist following discectomy, that in a far higher proportion the pain relief is medium term rather than permanent, and that a substantial proportion of patients with the symptoms and signs of a herniated intervertebral disc have no spinal pathology that can be detected by modern diagnostic techniques. In the absence of a disc prolapse, or as an alternative diagnosis, other pathologies are linked with back pain.

Degenerative arthritis, osteoarthritis, or osteoarthrosis (the multiplicity of names reflects the doubt over the pathology) affects the spine, as it does other joints with increasing age and is another unsustainable diagnostic label. With symptoms that are so common and a 'pathology' that is so universal, it is hardly surprising that the two have been linked. The synovial joints of the spine, the facetal joints, are supplied by branches from the nerves that form the roots of the sciatic nerve and the other peripheral nerves. Thus the concept of referred pain fits in well with the concept that arthritis in these joints might cause radiating back pain. Furthermore, the nature of the pain is all that one might associate with an arthritis rather than a nerve root compression.

It is possible that the origin of the pain might be muscular. Back pain is usually associated with tender muscle and muscle spasm. Pain arising in muscles supplied by a posterior primary ramus might be referred to the anterior primary ramus. Painful muscle and connective tissue, in the absence of other disease or injury, has been called a variety of names including fibrositis, muscular rheumatism, and fibromyalgia, and is dealt with elsewhere in this book. However, a substantial proportion of the back pain syndromes would fit in with the concept of myofascial pain.

Which patients recover from back pain and sciatica following which treatment—and does this give us any clue as to whether we are dealing with a variety of diagnoses or with a single or small number of 'diseases'? Perhaps, in a proportion of patients, the nuclear material from the ruptured disc can cause their symptoms. Nucleus pulposus is soft and might cause nerve compression, but low back pain is not the same as the pain of nerve compression elsewhere in the body. Carpal tunnel syndrome, ulnar nerve compression, and other nerve entrapments cause tingling dysaesthesia, as we would expect from our knowledge of pain perception. The eventual pain of chronic nerve compression would be of a deafferentation type. The pain of sciatica does not have these characteristics. It resembles the pain of a damaged muscle or joint rather than a neurogenic pain.

The pain of arthritis is difficult to understand and in joints elsewhere in the body the pain resulting from arthritis bears no relationship to the degree of arthritis present. However whenever we make the diagnosis of arthritic pain elsewhere arthritis is present. Arthritis is not invariably present in sufferers from back pain, even when they have all the characteristic symptoms that we might attribute to facet joint arthritis. Furthermore, treatments that might be expected to elicit a consistent response from arthritic joints do not produce a consistent response in back pain, and recovery can occur from treating non-arthritic joints as if they were arthritic.

Does this leave us with myofascial pain as the principal cause of back pain? Once again we are confronted with a failure of consistency. Patients with back pain and radiation in the distribution of peripheral nerves may have tender myofascial trigger points. When these are injected with local anaesthetic these patients may enjoy complete symptom relief. But tender muscle is not invariably present in low back pain and much of the symptomatology is not that of myofascial pain elsewhere in the body.

From this lengthy introduction it will be clear that the type of physical intervention that should be used in back pain must be controversial in view of the lack of certainty as to its cause. Patients with acute back pain and sciatica frequently recover within four weeks untreated and after one year, 85 per cent are symptom-free untreated. Is the right approach to admit ignorance and to desist from physical intervention? After all, all treatments have a cost and resources are limited and should be devoted to areas of medical care in which they can be used logically. Much of the treatment used for back pain is painful, and physical intervention will have a certain incidence of complication. There is a mortality associated with disc surgery. Bed rest for a short period may be the only type of medical intervention that might be justifiable. The withdrawal of active physical intervention might lower the development of chronicity. In the very small proportion of patients in whom back pain becomes chronic, psychologically based pain management techniques might be the most appropriate therapy.

Many types of therapy have been advocated for low back pain and sciatica and have been as enthusiastically adopted as surgery. These include thorough bed rest, traction, the tissue heating modalities of physiotherapy, manipulation according to the doctrines of a variety of schools (from osteopathy and chiropractic to 'mobilization'), acupuncture, trigger point injection, epidural injection, facet joint injection or denervation, and the relaxation, biofeedback, hypnosis, and cognitive methods of pain management of the non-interventionists. The advocates of each of these methods have a rationale for the technique that they favour. Each technique may relieve back pain effectively or its advocate would soon have no patients, so, to repeat the question that was asked above, less specifically, why do sufferers from back pain respond to treatment—and should one type of treatment or another be directed at all or some back pain sufferers?

Back pain causes distress and may be a sign of infectious or neoplastic disease; structural changes in the spine that have been labelled as a cause of back pain can be associated with progressive neurological damage. It can respond rapidly to physical intervention and there appears to be a specificity about the type of physical intervention that may produce relief in any individual incident of back pain. Where back pain is associated with objective evidence of neurological dysfunction, i.e. where a lesion can be found that might cause neurological damage and where this lesion corresponds with the pain that the patient is experiencing, then surgery is indicated to prevent further damage. In these circumstances the chances that the surgery will coincidentally relieve the pain are high. In young patients with structural abnormalities of the spine which are associated with pain in a distribution that would correspond with their lesion, surgical manœuvres to limit the movement of that part of the spine that may be giving rise to the symptoms will be very likely to relieve their pain. The further from these conditions that the indications for surgical intervention are removed, the less the chances of producing relief and the higher the chances that surgical intervention will be followed by chronic pain and disability.

Assessment

The assessment of the back pain sufferer in the Pain Clinic must start with the requirement to identify the patient with a lesion that should be corrected surgically. Is the symptom of recent onset or has there been a change in symptomatology recently? Is there evidence of decreasing function in the organs supplied by the cauda equina, such as perineal numbness, recent onset of difficulty in starting micturition, and rapidly increasing lower limb weakness? These symptoms and signs constitute a surgical emergency warranting immediate referral to a competent surgeon. Is there numbness or weakness? Severe pain radiating down a limb will give a subjective impression that there is a sensory change in the limb, and pain on physical use of the limb may give a feeling of weakness. There is no mistaking true numbness, and true weakness will give rise to a loss of power which will affect either extension of the thigh or dorsiflexion or plantar flexion of the foot, all of which will seriously interfere with gait. On physical examination, is there objective evidence that peripheral nerve function is disturbed? An anxious or hysterical patient will find altered pin prick and light touch sensation in the area in which pain is being experienced. Similarly, on testing for power, pain may interfere with the patient's ability to co-operate. Waddell *et al.* (1984) devised a series of tests to differentiate back pain with a physical basis from that in which the symptoms cannot be those that would be produced by physical disease of the spine and its contained and surrounding structures.

These are, in describing symptoms:

1. A pain drawing that shows the pain to be localized, neuroanatomical, and proportionate.
2. Pain adjectives that are sensory rather than affective and evaluative.

3. Pain described as localized rather than involving a whole limb or the tail-bone.
4. A pattern of numbness that is dermatomal and of weakness that is myotomal.
5. Pain that varies with time rather than being constantly present.
6. A response to treatment that varies in its benefit. Not an intolerance of treatments and emergency admissions to hospital with pain.

In physical signs:

1. Tenderness that is localized, not superficial and non-anatomical.
2. Lumbar pain is not produced by axial loading.
3. Straight leg raising should remain limited in spite of distraction.
4. Sensory changes should be dermatomal rather than regional
5. Weakness changes should be myotomal, not jerky and associated with 'giving way'.
6. The general response to physical examinaton should be appropriate rather than one of over reaction.

Patients should reach a Pain Clinic investigated, and back pain sufferers should have had a blood count and had their plasma viscosity measured to detect inflammatory disease, and an X-ray of the lumbar spine and pelvis to exclude a space-occupying lesion or trauma. A patient attending a clinic should be reassessed and reinvestigated whenever the symptoms change.

Management

The management of back pain in a Pain Clinic starts at the point where specific therapeutic manœuvres finish. Where this point may be, will vary with the point of view of the referring therapist but clearly, as diagnosis is so indefinite beyond the bounds of trauma and infectious or neoplastic disease, the Pain Clinic methods of back pain management should be resorted to before destructive physical invasion is considered. As diagnosis and pathology are so little understood it can be argued that the only logical way in which to manage patients with back pain is to progress through the various pain management techniques until one can be found that gives symptom relief. Thus, a logical progression is through the various therapies going from the treatment that is aimed at a variety of potential causes, to that targeted at more limited 'pathologies', and from treatments that involve no invasion or trauma to those that involve pain and discomfort, only destroying tissue where there seems to be excellent grounds for believing that, in so doing, the chances of producing useful symptom relief will be high.

Bed rest

This form of treatment is usually resorted to immediately. Acute back pain is itself immobilizing for several hours, as the pain of breathing or simple postural readjustment can be severe. There is no evidence that if the period of rest is prolonged beyond that enforced by the severity of the initial pain and muscle spasm there will be any reduction in the duration of severity of the attack. If the pain is caused by a lesion that is compressing a nerve root, then immobility might well reduce the nerve damage that is being caused. However, if the pain is the result of soft tissue damage, then prolonged immobilization will lead to muscle weakness, stiffness, slower resolution of any causative oedema, and disability.

Jackets and corsets

The same arguments will apply to the use of plaster jackets and corsets. As low back pain is exacerbated by movement and relieved, to some extent, by rest it seems logical to immobilize the back. Indeed, immobilizing techniques are likely to relieve the pain while they are applied. To this extent they are very valuable in the period immediately following the onset of pain, allowing the patient to return to a more active life and reducing invalidism and improving function. Immobilizing and supporting jackets and corsets should, however, be regarded as temporary measures as their continued use will lead to the usual problems of disuse. A reasonable parallel is the use of support for a sprained joint. It would be harmful to immobilize a sprained joint as the result would be prolonged stiffness, weakness, and disability. However, following a sprain the joint can be brought back to full function more rapidly if it is supported and its range of movement kept within pain-free bounds by some form of bandaging.

Manipulation

Pulling and twisting movements applied to the lumbar spine will instantly produce pain relief in many patients with back pain. This fact has been known from ancient times and provided a living for a variety of therapists. Because the medical profession has centred its ideas on drug therapy and surgical intervention, manipulation has been developed outside the field of conventional medicine. Practitioners of manipulation have developed theories as to its mode of action, and its use in the whole spectrum of disease as an alternative to conventional management or even to the practice of conventional medicine as a whole. No evidence has ever been produced that manipulative techniques have any benefit in the relief of disease arising outside the musculoskeletal system. The persistence of training for professions that believe that a multitude of diseases can be cured by manipulation, has done a great disservice to practitioners who have a lot to offer in the relief of back pain. It is perhaps also unfortunate that physiotherapy, chiropractic, and osteopathy, which possibly produce relief of back pain through a similar mechanism, have grown up as different schools, believing that each alone has the correct and professional answer to manipulative treatment.

The techniques involve putting the joints of the affected part of the spine (or when applied to a limb, the limb) forcibly through a full range of movement. The degree of force, the way in which it is applied, and the diagnostic techniques used before applying therapy vary between the different schools of thought. Chiropractic uses force directly applied to the vertebral column, osteopathy relies on traction and twisting manipulation, and the Maitland school of physiotherapy on progressively-applied, oscillating movements. One technique may work where another doesn't, but no technique has any superiority over any other in the presence of any particular group of symptoms or physical signs.

No manipulative technique should be used in the presence of objective signs of acute neurological damage such as loss of reflexes, perineal numbness and difficulty in micturition, or wasting. Although spinal cord and peripheral nerve are extremely tough, where the existence of nerve damage is already known, then clearly further stretching of the tissues around them could exacerbate such damage. We also believe that where there is radiological evidence of disc prolapse from a computerized axial tomography (CAT) or magnetic resonance imaging (MRI) scan, such prolapse can be increased by manipulation. There has never been evidence that manipulation reduces an existing disc prolapse; discs prolapse through the weakening and eventual collapse of the annulus fibrosus. Further torsion and stretching of the annulus cannot restore it to its original shape. It would clearly also be a disastrous error to manipulate in the presence of a space occupying lesion, intraspinal infection, metastases, or a fracture and these should be absolutely excluded before manipulation is undertaken.

A pain relief clinician should, however, be prepared to acknowledge that in the absence of these contraindications manipulation can produce rapid and complication-free pain relief. An effort should be made to identify the part of the spine from which the pain is arising, by identifying the dermatomal origin of referred pain, by paravertebral muscle spasm, or by localized spinal tenderness. The index finger is placed between the vertebral spine on each side of the affected root and the lumbar spine is flexed until these spines are their maximum distance apart. At this point the lumbar spine is rotated away from the side in which the symptoms have been felt. The degree of rotation is limited by the patient's tolerance, the manipulator's courage, and the school of manipulation that is being followed. Manipulation of the cervical spine follows roughly the same principles, except that the traction is produced by the manipulator's weight pulling on the patient's head using a flexed arm. Professionals make a whole career from manipulation and it can be argued that it is in the patient's interest that manipulation should be undertaken by a manipulator, in the same way that the care of the unconscious during surgery should be left to an anaesthetist.

The mechanism by which manipulation relieves back pain can only be a matter of speculation, as the cause of back pain is unknown and the effects of manipulation on bone, connective tissue, muscle, and nerve cannot be seen. Theories abound, for example: that manipulation 'puts back herniated discs' (which cannot be so), that it expands contracted scar tissue, that it stretches muscle in painful spasm, and that it moves trapped nerve (any of which is possible). It must suffice at present to understand that somewhat traumatic interventions that produce non-specific structural displacement will relieve acute and chronic back pain and that, up to a point, the more physically disruptive the manœuvre the more dramatic the relief.

Techniques using local anaesthesia

There is no more logic in the use of local anaesthetic nerve blocks than there is in the use of manipulation. That is to say that it is known, from the introduction of spinal and paraspinal nerve blockade, that spinal and epidural anaesthetics can relieve back pain. There are theories as to why this might be so that closely resemble those about the mechanism of action of manipulation. It must be reiterated that as we have no certain knowledge of the cause of back pain, so have we no knowledge of the way in which the techniques used to relieve back pain work.

Epidural injection

The argument as to whether to use this or other techniques in the relief of back pain must first depend on the balance between risk and benefit. Many studies have been made of the injection of local anaesthetic, with and without steroids, into the epidural and subarachnoid spaces. Most of these studies were unsound, being uncontrolled and without firm inclusion and output criteria. Controlled studies have shown no difference between epidural injection and other non-operative methods of back pain treatment. On the other hand, although the injection of steroid drugs into the subarachnoid space is associated with an unacceptable morbidity, the epidural injection of both local anaesthetic and steroid is remarkably safe.

Complications exist, in particular if the lumbar route is used, such as temporary (albeit extreme) hypotension, complete motor block resulting in respiratory arrest, and accidental dural puncture which occasionally results in a prolonged headache. These are unpleasant but if the injection is performed by a competent anaesthetist need not be life-threatening. If the sacral route is used, the success rate of epidural injection (but not necessarily of the relief of back pain) falls. However, the likelihood of all complications, other than local anaesthetic toxicity from inadvertent overdose or accidental intravenous injection, also decreases considerably.

Thus here, as with manipulation, we have a method of producing relief from back pain whose mechanism has no explanation, that is safe in skilled hands, that is unpredictably effective, but which may produce relief of pain when other methods have failed (Kepes and Duncalf 1985). No evidence has ever been produced that either the lumbar or sacral route will produce better pain relief in any particular circumstances and although the sacral route may be considerably more painful and has a higher failure rate as an epidural injection technique, the reduced incidence of minor but unpleasant complications means that it is to be preferred.

Subarachnoid injection of depot steroids has been associated with permanent neurological damage and is to be deprecated.

The use of steroid with the injectate produces better results than the use of local anaesthetic or saline alone. There is no difference in the benefit produced by any particular local anaesthetic, and the only advantage in the use of local anaesthetic over saline is that the injection of saline can be unpleasantly painful. No dose or volume of injectate has been shown to be ideal. it seems logical to use a volume of local anaesthetic that blocks the nerves serving the area in which the pain is felt, as this acts as a label demonstrating that the solution has reached the part of the spine from which the pain is arising. There is no advantage in using a long- rather than short-acting local anaesthetic except that with the shorter-acting agent the patient can leave the clinic more quickly. No study has been published showing an advantage in using repeated epidural injections over a single injection; however, we have encountered patients who benefit only afer a second or third injection, but never a patient who has been relieved by a fourth injection who was not relieved by a third.

Patients who have been treated by epidural injection should be followed up at four weeks, firstly because relief may be delayed for a week or more after the course, and secondly because very temporary relief is not clinically useful. Sufferers who are relieved of their pain for several months but whose pain then returns may be consistently relieved of their pain for a useful period by repeating the injection. If this results in epidural injections being repeated many times then a risk must arise that, even in the epidural space, the repeated use of depot steroids may give rise to tissue damage. It is therefore reasonable to use local anaesthetic only, after a fourth course of injections.

Where treatment involves physical intervention of the order of an epidural injection, controlled trials of the procedure using a placebo rather than the drugs are not possible. However, a dramatic treatment of this sort may produce a placebo response. This is not undesirable in itself as the sufferer is, after all, relieved of his pain, but a repeat of the procedure is unlikely to produce a similar result. Thus, when a patient who has responded once fails to respond on a second occasion it is reasonable to assume that a placebo response was obtained to the first treatment and that it would not be worthwhile trying further.

Facet joint and posterior compartment injections and lesions

The joints between the articular facets of the spine can become arthritic and this might be expected to give rise to pain. They are supplied by the posterior primary rami of the spinal nerves which, in the lumbar and first sacral segments, form the lumbar and sacral plexuses with their anterior primary rami. It was regarded as a reasonable assumption that pain arising from disease of these joints would be felt in the back and, when referred to the anterior primary rami, might radiate down the leg as sciatica. This impression appeared to be confirmed when manœuvres designed to denervate these joints generated pain relief in many patients. The early techniques involved using a fine blade but this was often followed by massive haematoma formation and so it was succeeded by denervation techniques using radio frequency lesion makers (Shealy 1976). In an attempt to improve results of denervation by first using local anaesthetic injections to predict the response, workers found that the response to injection of local anaesthetic, with or without steroid, into or around the facet joints, or on to the posterior primary ramus, not only relieved the pain, but did so for as long as the various nerve destruction techniques. In addition, Bogduk and Long (1980) have demonstrated that the nerve supply of the joint is complex and that it is unlikely that the various 'denervation' techniques have in fact achieved their object even when they have relieved pain. The indefinite nature of this response has led to a degree of cynicism concerning the rationale for the various techniques advocated and for the use of 'facet joint arthritis' as a diagnostic term. It is probably more logical to describe a pain that responds to injections into the facet joints and the nerves and muscles in their vicinity, as a part of the posterior compartment syndrome.

What, then, are the indications for managing back pain as 'posterior compartment syndrome'? The initial concept that the problem is the result of arthritis is worth pursuing until a better explanation is forthcoming. Back pain of this type would be expected to be felt in the low back with dermatomal radiation, but the evidence is that on stimulation of the structures around the facet joint, radiation may be very non-specific. Examination of the lumbar spine will show that the lumbar vertebral spines on each side of the suspected joint will be tender to pressure but there will be no abnormal neurological signs in the dermatomes in which the pain is being felt. Arthritic pain should be associated with stiffness and should decrease with exercise. The joints can be seen radiologically on an antero–posterior view of the lumbar spine but

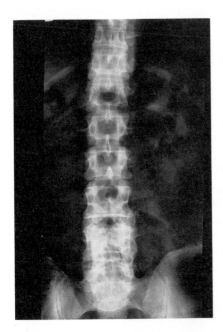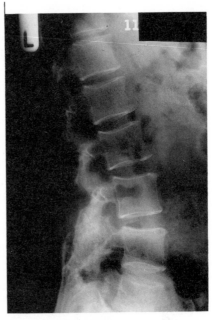

Fig. 4.1 Pain presumed to be arising in facet joints above a fusion of the fourth and fifth lumbar vertebrae. These facet joints would be subjected to abnormal stresses following the fusion.

can be better seen using rotated views. If arthritic, the suspected joints are narrow compared with the others and have sclerotic margins. It may be possible to see hypertrophy of the surrounding bone with osteophytes. The blood tests for the other arthritises should be negative.

Techniques

It is wrong to believe that any particular therapeutic regimen aimed at this area has any particular value over any other. In view of this, it is reasonable to first use the safest and least damaging method to hand, and only proceed to any method that involves tissue damage beyond that caused by the needle if there is good reason to believe that there is a good chance of producing more complete or long-lasting relief. Thus, injecting the joint itself, the tissues immediately around it, or the posterior primary ramus with local anaesthetic are the techniques that should be tried first. If the injection is to be intra-articular then it should be remembered that the joint space holds only 2 ml so injection volumes larger than this are likely to disrupt the joint and cause damage. Elsewhere in the body, intra-articular injections of steroid mixed with local anaesthetic relieve the pain of arthritis, so intra-articular injection of a depot steroid such as triamcinalone hexacetonide (*Lederspan*) is justifiable.

Because the joints are small, the injection should be radiologically controlled; the target is too small to hit 'blind'. If the patient experiences pain relief it is likely that this will last for several months but not indefinitely making it necessary to know precisely the site of the injection so that it can be accurately repeated.

If the injection is to be intra-articular the patient should first lie prone, and the joints be identified. Their surface position is marked using a radio-opaque object such as a large needle over the skin. As the joint is very small, a spinal needle (for example 25 or 26 SWG with an introducer) will be needed. There will be a feeling of 'give' as the joint is entered and resistance to injection may be high. It will be possible to see the needle enter the joint on the antero-posterior (AP) view; its position and the fact that it is in the joint and has not passed through into the intervertebral foramen can be checked on the lateral view. If resistance to injection is very high the needle is lying between the articular processes of a narrowed joint and should be slowly

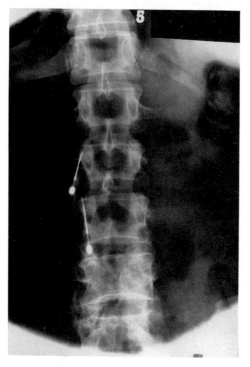

Fig. 4.2 Anterio-posterior view of X-rays taken during a facet joint injection. The upper needle is in the joint. The lower one will need to be moved medially.

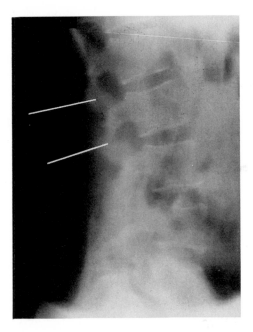

Fig. 4.3 Lateral view of X-ray taken during facet joint injection. The needle shadow has been enhanced. The lower needle was correctly placed in this view, although it was shown to be too lateral by the A/P view. It is essential to have both views.

withdrawn until injection becomes easier. The joint volume is so small that one should not inject contrast medium to verify the correct position of the needle tip; the whole injectate should consist of the local anaesthetic and steroid mixture.

If the injection is to be made into the tissues surrounding the joint, then once the needle has been positioned over the joint (AP view), the solution of local anaesthetic and steroid should be spread over the joint surface (lateral view). The posterior primary ramus can be blocked where it passes over the transverse process running from the intervertebral foramen below to the joint above, around the pedicle. The needle is inserted on to the transverse process as it meets the pedicle of the vertebra. Five ml of local anaesthetic will block the nerve. There is no need to add steroid to the local anaesthetic.

Response to these injections can be delayed; the patient should be warned that the injection can be followed by pain, but that any post-injection pain should have gone by six weeks. If the patient has useful pain relief by then, there is good reason to believe that it may be long-lasting and the injection can be repeated when the pain recurs.

If the response is definite and consistent but does not last more than a few days, then it is justifiable to create a heat lesion of the posterior primary ramus using a radiofrequency lesion maker. Its size depends on the temperature at the needle tip rather than the length of time for which it is switched on A lesion at 80°C in the groove between the pedicle and the transverse process of the vertebra below the affected joint will probably be accurate enough. Purists point out that the nerve supply to the joint is so complex that up to seven lesions would need to be made at specific sites in order to make a thorough denervation, but there is no evidence yet that this improves the clinical outcome.

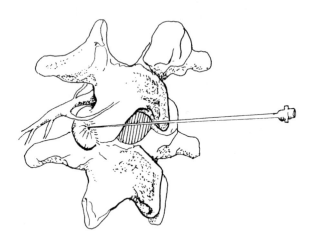

Fig. 4.4 Rotated view of vertebrae to show intra-articular injection into facet joint.

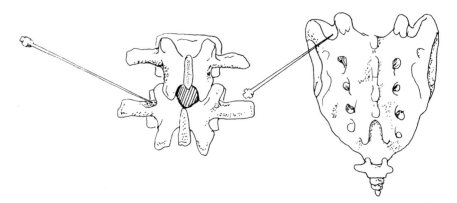

Fig. 4.5 Points for facet joint denervation (after Bogduk and Long).

Sympathetically maintained pain

Chronically maintained nociceptive stimulation may sensitize wide dynamic range neurons in the dorsal horn, so that they fire continuously whether or not the incoming stimulus is tissue-damaging. This firing is perceived as pain. One of the agents that will maintain this firing is peripherally released noradrenalin which acts on the receptors of these neurons or on sprouts growing from the axons of damaged or divided nerves. The pain felt is burning and may be associated with the other changes of sympathetic dystrophy. It will be relieved by sympathetic blockade. The firing of these neurons is caused by a change in adaptation and if the peripheral release of noradrenalin is greatly reduced even for a short period, the firing will stop and pain will be relieved for a much longer period. If the reduction in circulating level of noradrenalin is permanent, as a result of a sympathectomy, the pain will eventually return when the receptors become sensitized to this lower level of noradrenalin, which is still being released elsewhere in the body. This newly returned pain will not be susceptible to repeat sympathetic block as the sympathetic chain has already been destroyed and local release of noradrenalin is now extremely low and irreducible. If, however, a temporary sympathetic block is performed, the pain relief may last as long as that following a permanent sympathetic block, but be repeatable.

Myofascial pain and stimulation

Pain in the back might arise from any of the structures making up the spine, its contents, or the structures surrounding it; in bulk, muscle makes up the largest of these and in the absence of radiation into the leg, most back pain is felt in muscle. However, there is a common belief that back pain must be generated from within the bone or joint of the spine and muscle as a possible site is neglected. Indeed, although the sacrospinalis, the major posture-maintaining muscle of the back, arises over the posterior surface of the sacrum and injury or spasm in this muscle might be expected to be a major cause of back pain, pain arising from the region near the sacroiliac joint is almost invariably blamed on a problem in the joint rather than on this muscle.

This is not to say that there is any more evidence to implicate muscle rather than any other structure as the major cause of back pain. Although electro-myography (EMG) studies do not show any confirmable difference between the electrical activity of the posture muscles in chronic back pain sufferers and in others, if back pain arises from disorders of the structures of the back then it seems likely that at least some is generated within muscle.

Pain might arise from back muscle either from some disease such as polymyalgia rheumatica (whose diagnosis can be confirmed by raised blood viscosity, a positive Rose–Waaler Test, and biopsy), trauma, or the development of 'myofascial' pain. Back muscle may be strained just as may any other muscle in the body; following stress which is greater than the muscle can withstand, it become swollen and intensely painful. The pain usually reduces steadily over the course of a few days and resolves but muscle injury can become chronic if a part of the muscle bundle remains permanently altered and shortened with an increased electrical excitability. On electron microscopy, this area of muscle shows changes and it can also be palpated as a painful nodule. The shortened band of muscle fibre on each side of it, which can be palpated, can be straightened if the area is cooled with a cold spray (such as an ethyl chloride spray) or if the nodule or 'trigger point' is injected with local anaesthetic. It can sometimes even be relieved if the tender spot is needled with a dry needle and nothing injected at all. It is possible that a great deal of the success attributed to acupuncture results from this.

Acupuncture

Needling of tissue results in the release of humoral substances such as bradykinins, substance P, and leukotrienes because of tissue damage. It also results in the release of encephalins in the dorsal horn of the spinal cord. If the needle is inserted into a myofascial trigger point the muscle will cease to be painful and if the needle is inserted into the tender points that surround an arthritic joint, that joint will become less painful. These symptomatic changes are progressive.

Studying an invasive procedure such as acupuncture and attempting to show that any effect is specific to the treatment rather than a placebo response is extremely difficult. The effect of inserting a needle can be felt by the patient and so cannot be compared on a 'blind' basis with not inserting a needle. Operator bias cannot be excluded either, since an acupuncturist cannot administer what he believes to be real rather than false acupuncture without being aware of the type he is using. Extensive studies have not revealed any particular merit of one school of acupuncture, such as traditional acupuncture, over any other, such as dermatomal or trigger point acupuncture, in achieving pain relief. All that is needed is to insert the needle and to manipulate it until an aching sensation is felt. Often this manipulation will produce a flush around the needle, indicating the release of tissue-damage humoral factors. In view of this lack of specificity, it seems reasonable to needle trigger points in muscle and the tender points around joints and in the area where pain is felt rather than attempting to use 'traditional' points which apparently have no relationship to any rational approach to treatment. An acupuncture-like effect can be produced by increasing the output and decreasing the frequency of transcutaneous stimulation so that muscle contraction is produced. This low frequency, high intensity stimulation has been shown to produce pain relief when low intensity, high frequency stimulation has failed.

Low threshold, high frequency transcutaneous stimulation to produce pain relief is aimed at inducing inhibition in the dorsal horn by stimulating the fast-conducting nerve fibres which subserve light-touch receptors. If it is to work there must be functioning normal light touch receptors and fibres in the area being stimulated. These receptors should be as close as possible to those nociceptors and their associated fibres which are suspected of carrying the pain centrally. If an area is stimulated in which there is abnormal sensation and light touch sensation is distorted, this sort of low intensity generated inhibition is unlikely to be effective. Similarly, if the stimulation experienced is felt as unpleasant, it is probable that nociceptors are being stimulated and that the onward transmission of pain information, far from being inhibited, will be facilitated. It should also be borne in mind that low threshold receptors accommodate if the frequency of stimultion is too high.

Before transcutaneous electrical nerve stimulation (TENS) is used, the light touch sensation over the area should be checked. Where it is disturbed the electrodes should be applied as close to the disturbed area as is practical. The pulse width and pulse height should be adjusted until the stimulation is felt to be uncomfortably intense, then the intensity should be decreased until it has just become entirely pleasant. There should be no burning or pricking elements to the sensation and if there is, the frequency should be adjusted until it is just below that at which this sensation starts to fade after a few minutes. The stimulation must be felt all the time.

There is evidence to suggest that stimulation produces the same amount of pain relief as distraction by another stimulus of the same intensity. From this it could be inferred that it is the distraction of the stimulus which generates descending inhibition rather than some separate inhibitory effect specifically generated at the dorsal horn. There is also experimental evidence, however, that the analgesia is produced at the dorsal horn, and until more specific evidence becomes available every effort should be made to make use of the theoretical basis for low intensity stimulation.

Drugs

Sufferers from pain, including back pain, wish to be 'cured' by a course of treatment that ends with long-term relief from pain without further treatment. Where this cannot be achieved using drugs then it is better that some peripheral technique such as TENS is used to generate long-term pain relief. The use of any drug involves dangers and effects other than pain relief.

Non-steroidal anti-inflammatory drugs

These drugs inhibit the enzyme cyclo-oxygenase, preventing the conversion of arachidonic acid, released as a result of cell membrane damage, to prostaglandins. There are considerable doubts as to whether this action of these

drugs is responsible for all their analgesic effect since the analgesia is not proportional to their potency as cyclo-oxygenase inhibitors and there are other
products of cell damage, such as leukotrienes, responsible for pain whose
production is not affected by the NSAI drugs. Furthermore, paracetamol,
one of the most effective drugs in this group, has very little effect on prostaglandin synthesis. It may well be that the beneficial effect of all these drugs
arises from a non-specific analgesic action. Their use is associated with
several side-effects, whose incidence increases with the patient's age. As a
result of the inhibition of prostaglandin synthesis, the production of the protective mucus barrier on the stomach wall is inhibited and a resulting increase
in gastrointestinal blood loss is invariable. The danger of major gastrointestinal haemorrhage is increased by an associated reduction in platelet aggrega-
tion. These drugs can also induce a disease of the lower bowel similar to
Crohn's disease. In older subjects, a group in whom the incidence of back
pain is greatest, NSAI drugs become increasingly hepato- and nephrotoxic.

For all these disadvantages the rational and careful use of these drugs can
provide a short- to medium-term solution to the back pain problem. Sensitivity to, and tolerance of different drugs in the group will vary from patient to
patient. The pain clinician should have a group that he is used to handling,
and before any drug is rejected it should have been taken at the maximum
dose recommended, or in as high a dose as the patient can tolerate, for long
enough for its effect to have been assessed. In particular this applies to paracetamol, which is extremely safe taken at the recommended dose. When
treating continuous pain, whatever the cause, the principle applies that to be
effective the drug must be taken at such a dose and such a frequency that pain
is not experienced. If the pain relief is incomplete or the pain is experienced
between doses the patient will continue to suffer and require increasing doses
or more powerful analgesics.

Opioids

So-called mild opioids are so commonly prescribed in back pain that it would
be unusual to see a patient with back pain in a Pain Clinic who was not taking
them or who had not rejected them as inadequate or been unable to tolerate
the side-effects. As the difference between the mild and powerful opioids is
one only of degree, there is no special merit in the use of dihydrocodeine or
dextropropoxyphene over an equipotent dose of methadone or morphine.
The hazards of addiction and tolerance remain the same. The first question to
be answered if the use of opioids is to be considered is 'Is this back pain opioid
sensitive?'. Various tests have been described, for example testing whether the
patient can distinguish between opioid and saline injected intrathecally, or
intravenously. The most specific of these is an intrathecal test.

If an opioid is to be used, the same principle applies as that described for
NSAIs: the drug used should be taken at such a dose and at such a frequency

that no pain is felt. The weaker opioids are often combined with paracetamol and an attempt to reach a high enough opioid intake to give complete pain relief would mean taking paracetamol in poisonously large doses. Thus tolerance and dependence often develop. Codeine, dihydrocodeine, and dextropropoxyphene are really not powerful or long-acting enough to give full analgesia in severe pain and if the dose is pushed above the recommended maximum in an attempt to reach complete analgesia, unpleasant side-effects such as hallucinations and severe constipation result. If opioids are to be used then, as in other diseases, morphine probably represents the best choice. The morphine should be taken 4-hourly as an elixir in progressively increasing doses until the pain is not felt. This total 24-hour requirement should then be converted into slow release morphine to be taken twice daily.

Antidepressants

Analgesics inhibit perception of nocigenic pain. They do not interfere with the perception of pain arising from any point proximal to the receptor in the peripheral nerve. Peripheral nerve can be damaged by the bony changes around diseased vertebrae, by compression within narrowed foraminae, or by surgery. The cord can be similarly damaged by changes in the spine, by surgery, by diseases such as spinal tumours or epidural abcesses, or by trauma. Back pain need not be associated with any detectable tissue damage, being generated by a defect of perception at cortical level. Where the pain is neurogenic rather than nocigenic it will not respond to analgesic drugs, but other drugs that stabilize the nervous system and alter perception may be effective. Of these, antidepressants are the most studied. Their action may be at a variety of sites. There is some dispute about whether they are analgesic, but the balance is in favour of their being so. They affect a variety of nervous system functions as well as their effect on mood for which they are primarily used therapeutically. Even if one were to depend on this action alone, it is possible to understand how they might relieve chronic pain. Anxiety and depression play a central part in the chronic pain syndrome and reduction of these emotions will be clinically beneficial. For this reason the tricyclic drugs such as amitriptyline, which produce sedation as well as relief from depression, can be thought of as more logical than the use of other antidepressants with less sedative qualities. Very much smaller doses of these drugs will be needed in pain relief than in depression; it is advisable to start with the smallest possible dose and build up until pain is relieved or side-effects become intolerable.

Anticonvulsants

The use of anticonvulsants in the relief of neurogenic pain has developed from their successful therapeutic effect in trigeminal neuralgia. Anecdotally they are thought to be most useful where pain is stabbing in nature. Their use is widespread and although they are almost certainly effective there is as yet no evidence from properly controlled trials that they are more effective than placebo. In a little understood condition such as chronic back pain the use of drugs with no proven value and their own risks should be weighed carefully against any possible benefit.

Pain management

As in other chronic pain conditions, chronic back pain generates invalidity. The sufferer will reduce his working and social life because to continue them at a healthy level hurts and the pain creates an incorrect belief that damage is being done. The reduction in activity will not reduce the amount of pain being felt and will be quite valueless. If the patient is encouraged by his family, the medical profession, and his friends to adopt the role of an invalid, this useless and harmful behaviour is reinforced.

Using psychologically based techniques the patient can be taught that this reduction in activity is valueless and that the back pain need not prevent full involvement in life. These pain management methods are described elsewhere in this book.

References

Bogduk, N. and Long, D. M. (1980). Percutaneous medial branch neurotomy. A modification of facet denervation. *Spine*, **5**, 193–290.

Kepes, E. R. and Duncalf, D. (1985). Treatment of backache with spinal injections of local anaesthetics, spinal and systemic steroids. A review. *Pain*, **22**, 33–48.

Mixter, W. J. and Barr, J. S. (1934). Rupture of the intervertebral disc with involvement of the spinal canal. *New England Journal of Medicine*, **211** (5), 210–15.

Mooney, V. and Robertson, J. (1976). The facet syndrome. *Clinical Orthopaedics*, **115**, 149–56.

Shealy, C. N. (1976). Facet denervation in the management of back and sciatic pain. *Clinical Orthopaedics*, **115**, 157–64.

Waddell, G., Bircher, M., Finlayson, D., and Main, C. J. (1984). Symptoms and signs: physical disease or illness behaviour? *British Medical Journal*, **289**, 739–41.

5 Body wall and myofascial pain

Pain in the body wall may come from the skeleton, the muscle and connective tissue of the body wall, or as the result of peripheral nerves becoming trapped in these tissues.

Skeletal pain

Pain in the body wall may be referred from the spine or arising from the ribs. Although referral patterns may often be very precisely distributed in a dermatomal pattern, studies of pain generated by stimulation of the spinal facet joints has shown that this is by no means always the case; for example, stimulation of the L5/S1, L4/5, or L3/4 facet joints may all cause groin pain. The management of back pain is discussed in Chapter 4. If the cause of body wall pain cannot be found elsewhere it is worth considering the possibility of a spinal origin.

Abnormalities of the thoracic cage may cause pain. A slipping or clicking rib syndrome, a pain at the costal margin aggravated by movement which may arise from some abnormality of the costal cartilage, is described. The symptoms, signs, and management are the same as those for an entrapment of one of the middle intercostal nerves and meticulous examination will usually reveal this as the cause of localized tenderness with radiating pain in this area.

Myofascial pain

Chronic pain arising in muscle associated with trigger points and tender, taut bands is a common cause of body wall pain. It is one of the chronic musculo-skeletal pain syndromes and although for many years diagnostic terms used to describe pains of this sort were vague, there is now a sufficient number of detailed studies of the various varieties of pain of this nature to make it possible to divide the myofascial and other chronic musculoskeletal pains into specific syndromes.

Fibromyalgia

In primary fibromyalgia (PFM), there are multiple tender spots in various muscles of the type formerly described as fibrositis. The syndrome is also characterized by generalized muscle aching, tiredness, and a disturbed sleep pattern. The concept that primary fibromyalgia can be identified as a

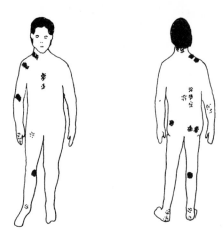

Fig. 5.1 Sites of tenderness in fibromyalgia. The areas in which shading is densest indicate the most common sites.

syndrome is so recent that strict criteria are still a matter of debate, but a synthesis of the suggested criteria might be that there should be widespread aching of more than three months duration, many areas of localized muscle tenderness, and sleep disturbance. There must be no objective evidence of muscle or joint disease. The tender spots may be found all over the body but are consistently grouped over the belly of the sternomastoid, over the extensor origin of the forearm muscles, down the thoracolumbar spine in the sacrospinal is muscle, and in the upper part of the gastrocnemius. These points remain constant for the individual patient. They are not 'trigger points' of the type found in the myofascial pain syndrome; a description of this syndrome will follow. There have been descriptions of changes in affected muscle when seen under electron microscopy, but there is still no clear evidence that these are specific to primary fibromyalgia.

The changes in sleep pattern are much more specific and objective. Patients suffering from primary fibromyalgia awaken from a complete night's sleep feeling tired and unrefreshed. They have no problem in getting to sleep and do not wake during the night. They have an altered ratio between dreaming, rapid eye movement (REM) sleep and non-dreaming sleep where there is no rapid eye movement (non-REM). While they are in non-REM sleep, during which there should be a delta rhythm on the electroencephalograph (EEC), an alpha rhythm intrudes. The occurrence and incidence of this alpha rhythm is consistently related to the number and tenderness of the tender muscle points. If normal people with no myofascial tender points are aroused during non-REM sleep they will develop a myofascial syndrome.

Fibromyalgia may develop in patients suffering from rheumatoid arthritis, systemic lupus erythematosus, and hypothyroidism. This is secondary fibromyalgia. In hypothyroidism the fibromyalgia responds to thyroid replacement.

A wide variety of therapies has been directed at fibromyalgia but few have been subjected to proper scientific assessment. Three treatments have been shown to be superior to placebo; amitriptyline, cyclobenzaprine, and strenuous physical exercise. Cyclobenzaprine is not available in Britain. Amitriptyline alters sleep patterns, increasing the proportion of non-REM to REM sleep. This begs the question as to the cause of primary fibromyalgia. Are these anxious, depressed people whose sleep pattern becomes disturbed in such a way that they rouse themselves from non-REM sleep? Their muscles never undergo a deep relaxation, and as a result become chronically exhausted and stiff. Once this pattern is broken and complete relaxation is restored their muscles recover and their pain is relieved. On the other hand, is the condition one in which the muscle is abnormal and painful? This pain arouses the sufferer from non-REM sleep repeatedly and only when the patient can benefit from the complete relaxation that amitriptyline-induced, non-REM sleep affords can the muscle heal.

The fitness programme that has been used for treating PFM might work on the basis that complete muscle relaxation following an intense period of muscle metabolism will relieve the pain. The patients in the trial exercised through a full cardiovascular fitness programme. They did much better than a group of patients who had flexibility exercises.

Myofascial pain syndrome

While PFM sufferers have generalized muscle pain and stiffness with multiple tender points, together with an altered sleep pattern and stiffness, in the myofascial pain syndrome (MFS) pain is localized. There is often only one or very few tender points, which are trigger points. The trigger point lies in a band of tight muscle and if this muscle is pinched or pressed it twitches and referred pain is felt at a distance from the trigger point. The aetiology of MFS is as debatable as that of PFM. Possibly a localized point in the muscle becomes shortened as a result of prolonged fatigue and trauma. This point remains permanently shortened and hypercontractile. As in PFM no pathology has been consistently found in affected muscle, however there is an increase in EMG activity. Furthermore the tenderness can be measured with a calibrated pressure-inducing device, the dolorimeter, and a decrease in pressure sensitivity can be demonstrated as recovery develops.

Simons and Travell (1983) have made a study of the referred pain that arises from the trigger points in many muscles and have derived a treatment system based on stretching these trigger points. Once the point has been identified it is anaesthetized by injecting a small amount of local anaesthetic or, alternatively, the taut muscle bundle is sprayed with a cold spray. The muscle is then stretched passively as far as possible. This treatment must be frequently repeated until the condition resolves.

The temporomandibular joint and the muscles surrounding it are vulnerable to myofascial pain. However, because they come within the sphere of dentistry, pains in this area have been classified into a different group, the temporomandibular pain and dysfunction syndrome. The joint is subjected to great stress in chewing and can be damaged, and can also develop arthritis and meniscal damage. These are not the province of a Pain Clinic, and where there is a possibility that they are the cause of pain, the sufferer should be referred to an oral surgeon. The sufferer should be treated appropriately when the symptoms of PFM or MPS are present, that is, multiple tender points in muscle elsewhere in the body, together with tiredness, stiffness, and sleep disturbance, or a trigger point in the masseter or the pterygoids with referred pain.

The diagnosis of PFM or myofascial pain should be borne in mind in neck and shoulder pain for which no skeletal or neurological cause can be found. Whiplash injury, severe and continuing pain following extreme forced extension of the neck, in particular following a shunt car crash, is a particularly severe form of pain of this type. The trapezius, sternomastoid, and scalene muscles play as prominent a role in posture as the large muscles of the back, and minor unhealed injuries in them can give rise to severe pain related to movement with radiation down the arms as far as the hands. The patient may believe there are sensory changes and weakness in the hands, but on neurological examination no changes can be found. On examination the muscles supporting the pectoral girdle will be found to have many areas of tenderness, either true trigger points or tender spots. Treatment by the methods described above is frequently disappointing in this area, and possibly following the more severe neck strains the damage is so extensive and the resulting pain and spasm so severe that, in spite of exercise, stretching, and amitriptyline, the condition becomes self-perpetuating. There is evidence that early mobilization, exercising to the full limit of pain-free movement, produces far better results than the more traditional method of dealing with such injuries by rest and analgesia.

Nerve entrapment

Trapped nerves are blamed for many pains in the absence of real evidence that this is what is really occurring. When a nerve is trapped it will be compressed and will not function normally. There will be loss of power in the muscles it supplies, loss of associated reflexes, and muscle wasting. On the sensory side there will be alteration of light touch perception and blunting of pin prick sensation. There will be pain if the nerve is damaged or is moved violently within the entrapment.

In the body wall, nerves pass through fascial tunnels in muscle to reach and innervate the surface. These are found at the anterior ends of the intercostal spaces and at the lateral border of the rectus sheath. These nerves may become entrapped as a result of a change in body shape, such as follows obesity or pregnancy, as a result of post-operative scarring, after sudden violent twisting movements of the trunk such as in dancing, or as a result of prolapse of a small lipoma into the hiatus in the dense fascia through which the nerve passes.

Pain from a nerve entrapment may be severe. It will be related to movement, and relieved in certain positions. It may mimic visceral pain, but there will be no symptoms of a visceral disorder. On examination, tenderness will be limited to a very localized, finger-tip-sized area over a point where a peripheral nerve is known to pass through dense fascia in the body wall. The diagnosis can be confirmed if an intradermal wheal of local anaesthetic is raised over the painful point. A needle is inserted through this wheal and if a nerve entrapment is present the sufferer's pain can be reproduced by the needle. If a small volume of local anaesthetic is now injected the pain is abolished almost immediately. The pain may be permanently relieved by injecting some local anaesthetic and long acting steroid, perhaps by necrosing prolapsed fat or relieving neuritis. If this is unsuccessful then repeated small injections of 6 per cent aqueous phenol at monthly intervals should be used or, alternatively, a radiofrequency lesion made. As in other nerve damage pains, there is no treatment that will invariably result in pain relief.

References

McCain, G. A. and Scudds, R. A. (1988). The concept of primary fibromyalgia (fibrositis): clinical value, relation and significance to other chronic musculoskeletal pain syndromes. *Pain*, **33**, 273–87.

Travell, J. G. and Simons, D. G. (1983). Myofascial pain and dysfunction. The trigger point manual. Williams and Wilcox, Baltimore.

Wolfe, E., Hawley, D. J., Cathey, M. A., Caro, X., and Russell, I. J. (1986). Fibrositis: symptom frequency and criteria for diagnosis. *Journal of Rheumatology*, **12**, 1159–63.

6 Relief of pain in malignant disease

The relief of pain and distressing symptoms in malignant disease is often given the title of terminal care. This is unfortunate as it implies that control of symptoms is only appropriate when there is no further scope for active treatment. Symptomatic management of disease should begin whenever symptoms become a cause of distress or discomfort to the patient, and can continue alongside palliative measures aimed at slowing the progress of the disease. Although the disease may reach a stage at which treatment aimed at the disease process is no longer in the patient's interest, symptom control should be continued as long as life continues. This process encompasses not only management of the patient's physical symptoms, but also consideration of the social, domestic, and emotional consequences of the disease. The aim should be to help the patient to approach the end of life with minimal distress for him or herself, and maximal support for their carers. Although treatment is not aimed at altering the course of the disease, some syptomatic treatments may temporarily alter the course of events.

Surprisingly, perhaps, the concept of symptomatic control is relatively recent, or at least, recently rediscovered. Symptomatic treatment of pain took at least equal place with cure-orientated management for previous generations, but the situation changed with the development of scientific medicine in the latter half of the nineteenth century, when investigation and aggressive treatment of the disease took precedent over managing the patient's symptoms.

The development of the hospice movement associated with such pioneers as Dame Cicely Saunders over the past thirty years or so, not only provided the physical environment in which to manage symptom control, but also had a powerful effect in gradually changing the attitudes of the medical profession. The all too prevalent attitude that medical science had nothing further to offer gradually began to give way to the concept of appropriately controlling those symptoms which distressed the patient, as opposed to those who predominantly concerned the doctor. Perhaps partly because of prevailing medical attitudes, this approach required a physically different environment in which to care for the patient with malignant disease, but with better training of medical and nursing staff, it is now increasingly possible to care for patients with distressing symptoms both in the community and in the hospital setting. Hospices have provided valuable research and experience in the effective use of opioid analgesics in chronic pain. As their use has gained wider acceptance, in countries where opioids are readily available for medical use the more invasive and destructive procedures for pain relief have become less necessary. Sadly, this is not universally true, and in many regions of the world destructive methods of alleviating pain are all that can be used, assuming that the necessary skills are available.

As well as analgesia and symptom control, the hospice provides an environment which is more acceptable to relatives during the late stages of malignant disease than is an acute hospital. Whilst expert medical care is available, with perhaps more individual attention than is possible in a general hospital, the environment also permits more spiritual care involving a close personal relationship with the nursing and medical staff, as well as support from professional counsellors, social workers, or clergy. Patients may find it easier to come to terms with their condition in this atmosphere than in the rushed, disease-orientated hospital or in the often emotionally fraught domestic situation. This does not necessarily require permanent residence in the hospice, and many patients are managed in day units or are admitted for a short period for symptom control or family respite before continuing their care at home.

The hospice acts as an important educational resource, advising and training medical and nursing staff and other health care professionals, as well as acting as a base for community nursing services. McMillan nurses often work from the hospice to provide continuity of care between hospital, GP, and hospice.

Another development has been the concept of the hospice in the community. This avoids the need for a costly building and provides a team of medical and nursing staff to provide specialist care for patients in their own homes, as well as support and advice for general practitioner services.

Incidence of pain in malignant disease

Analysis of published reviews (Bonica 1985) reveals that up to 70 per cent of cancer patients suffer from pain as a symptom, and in at least 50 per cent of patients undergoing treatment for cancer, pain continues to be a problem (Foley 1979). In 1978 there were 126 788 deaths from cancer, which means that about 42 000 patients had inadequate pain relief, and several surveys conducted recently in British hospitals suggest that these figures have not substantially improved, despite the increased attention given to this area.

Assessment

The assessment of the patient with malignant disease requiring symptom control is complex and requires the same skills as those used in medically assessing any patient. However, patients attending a Pain Clinic have usually been extensively investigated in relation to the pathological nature of their disease and have received treatment aimed at altering the progression of their disease. If we are being requested to provide treatment aimed at relieving symptoms, it is essential to obtain as full a history as possible of previous treatments and the results thereof. We need to know what, if any, further treatment of the disease is planned and what is the current medication and its related side effects. What is the diurnal variation of the pain, and does it limit movement or sleep? How is the patient being cared for at home or in hospital, and what effect is this having on the carers?

Pain in the patient with malignant disease may be a direct result of that disease, or it may be coincidental. The pain may be unrelated to the cancer and be due, for example, to long-standing musculoskeletal problems. It may result from general debility, such as bedsores, or it may be as a result of treatment; surgery, chemotherapy, and radiotherapy can all produce painful inflammatory changes.

It is important to try to decide on what is the cause of pain, as this will influence treatment. Is it due to visceral obstruction or distension? Is there an element of tissue destruction or inflammation contributing to the pain? Is pain resulting from compression or invasion of nerve tissue? Is the pain resulting from, or aggravated by, mental pain and anguish in the patient either at the prospect of impending decline or due to distressing symptoms other than the pain? It is obviously essential to enquire about domestic circumstances and the incidence of other symptoms.

Goals in pain management

The aims and expectations of pain control should be realistically determined and agreed with the patient. Unrealistic goals will result in disappointment and damage to the mutually trusting relationship between doctor and patient. Complete pain relief is not always possible, but improvement in the quality of life is possible in the majority of cases.

The first goal, and that usually most readily agreed by patients, is to achieve a restful night's sleep, undisturbed by pain. Fortunately, this goal is frequently the easiest to achieve. The next goal is to obtain acceptable daytime pain relief at rest and then, perhaps the most difficult of all, is to obtain pain control during activity and hence to improve mobility and independence. It is better to discuss at an early stage how much you hope to achieve, rather than to allow the patient to expect total relief. Many patients will accept a compromise between complete pain relief and undesirable side effects of treatment, and frequently gain tremendous help from the knowledge that someone is concerned with improving their quality of life and relieving some of their distressing symptoms. Reassurance is a powerful placebo.

Table 6.1 Pain in malignant disease: a summary of pain syndromes and their management.

System	Problems	Initial treatment	Also considered
vertebrae	aching in spine, sharp pain in spine or peripheral nerve distribution, weakness, numbness, bladder disorder	opioid with NSAID, radiotherapy, surgical decompression, immobilization	nerve block, hormonal treatment
long bones	aching, pathological fracture with pain on movement	surgical fixation, radiotherapy, analgesics	hormonal, direct injection
ribs	localized pain or fracture	NSAIDS, nerve block	dorsal root ganglion lesion
skull	headache, tenderness, symptoms of raised intracranial pressure	steroids	analgesics
visceral	dull aching pain, dysfunction of affected organs, pain referred to body wall, shoulders, perineum, etc.	opioids, steroids if nerve compression or liver distension	coeliac plexus or lumbar sympathetic blockade
nerves	pain, paraesthesiae, hypersensitivity, numbness, weakness, autonomic dysfunction	steroids for nerve compression. tricyclics for burning or dysaesthesia; anticonvulsants for shooting pain. sympathetic block.	radiotherapy, nerve blocks, TNS
soft tissue	local pain, tenderness	analgesics, NSAIDs	treat infection, nerve block
bowel	colicky abdominal pain	treat constipation, surgery, antispasmodic, opioid	
	tenesmus	chlorpromazine, nerve blocks	lumbar sympathetic block

Principles of analgesic management

Drug therapy is the principal method of controlling pain in malignant disease. Other methods, including invasive and destructive techniques, are nearly always a second choice. Drug treatment is generally reversible, titratable against the symptom, and introduces an element of patient involvement which may have valuable psychological benefits.

If a pharmacological approach is determined, then the first choice for route of administration is by mouth. Only if this is impracticable or ineffective are other methods considered. The patient must be constantly reassessed, so that rapid adjustment of dose or change of drug can be made according to the response, and for this reason the active involvement of appropriately trained and dedicated nursing staff is invaluable. The dose of analgesic is individually determined for each patient and formulary guides to dose range can only be accepted as a guide to useful starting doses. The correct dose for any patient is that dose which gives maximum relief with the minimum of side-effects. The advantage of morphine over weaker opioids and partial agonist opioids is that morphine can be given in increasing dose to produce greater analgesic effect (provided that the pain is opioid sensitive), whereas the effect of the latter rapidly reaches a plateau, when increasing the dose merely increases the side-effects.

Pain is often classified as nociceptive, neurogenic, or psychogenic. Much cancer pain comes into the first category, some into the second, and emotional responses to the disease process can certainly influence both types. This has important implications for the choice of drugs used to manage pain. Opioid drugs generally are most effective against nociceptive pain, although their psychogenic properties are important. Anti-inflammatory drugs and simple analgesics are again primarily of use in controlling nociceptive pain, whereas pain with a neurogenic component is frequently unresponsive to classical analgesic drugs and its relief may require the use of drugs which are not usually classified as analgesics.

Analgesics should be used in such a way as to produce a constant effect rather than intermittent relief, but patients are often seen whose pain is inadequately controlled because they are taking a drug with a short half-life, too infrequently. Pain returns intermittently, along with the other symptoms of opioid withdrawal, resulting in an intermittently distressed patient who constantly watches the clock waiting for the next dose. This phenomenon is almost certainly partly responsible for the unjustified reputation attached to opioid drugs, concerning rapid development of tolerance and unpleasant addictive behaviour. These features are rarely a problem when appropriate drugs and correct dose regimens are prescribed. There is some evidence that analgesic drugs are more effective if administered in anticipation of pain rather than after the full development of the symptom, and experience tends to confirm this (McQuay *et al.* 1988). Another advantage of maintaining a constant plasma level of analgesic drug is that by avoiding the peaks and troughs of drug concentrations, the side-effects accompanying sudden surges in plasma levels will also be avoided. For these reasons, a patient with chronic pain who is managed with opioid analgesics should have the analgesic prescribed to be given by the clock and at time intervals which are known to be compatible with the effective period of action of that drug. There is no place for *pro re nata* medication in treating chronic pain.

There is no correct stage in the development of malignant disease when opioids should be started. There is certainly no justification in reserving opioids for the final stages of the disease. A reason often given for the latter approach is the fear that tolerance will develop and that the drug will no longer be effective when the need is greatest. It has been shown that patients can be managed on a constant dose of opioid for long periods once an adequate dose has been achieved to control pain. A patient with inadequately controlled pain may well continue to demand higher doses of analgesic. Once a stable dose has been achieved, any increased requirement often reflects a progression in the disease which requires a new stable level of analgesic to be determined. Tolerance to opioid analgesia may develop slowly, but generally reaches a plateau after about three months at a dose which may then be stable for a long period (Twycross 1974). Tolerance to any drowsiness often develops after a few days, and to nausea over a couple of weeks. After this time it may not be necessary to continue with routine anti-emetics although, of course, these may be required to control nausea resulting from the disease process.

Addiction is an emotive word and has little real meaning in the therapeutic context. Drug-seeking behaviour may be a feature of inadequate analgesia or of short-acting, euphoria-producing drugs administered with inadequate frequency. Drugs taken in direct response to pain do not tend to produce abnormal drug-seeking behaviour, especially when longer-acting drugs are used to produce a constant plasma level and therefore avoid the euphoria resulting from rapid peaks in plasma opioid levels. Patients taking opioids for a demonstrably opioid-sensitive pain rarely exhibit the behavioural abnormalities of those who self administer drugs for their psychogenic properties. Some physical dependence may develop in response to long-term analgesic therapy, but this is not a problem in the patient with malignant disease, as they will usually continue to require drugs for the span of their life; in those occasional situations where an alternative method of pain relief is instituted (such as a nerve block), it is relatively easy to wean the patient off the opioid drug over a short period of time (Walsh 1984). In this situation some patients are able to cease opioid medication quickly and with few side-effects (Evans 1981). Interestingly, cases have been reported in which a patient who has been managed for a long period on large doses of opioid has had a pain-relieving surgical procedure performed but has continued on their previous dose of opioid. This has resulted in the development of severe opioid toxicity requiring respiratory support and opioid-antagonist treatment. It appears that chronic pain acts as a natural antagonist to many of the usual pharmacological effects of opioids.

An analgesic staircase has been described (Fig. 6.1) where analgesic drugs are grouped together with other drugs of similar potency. This concept emphasizes the point of not administering additional drugs of similar potency if one drug from that group has been inadequate to control pain. Although this is in many ways a sound concept, it is also important to consider the benefits that may result from prescribing simultaneously two drugs which alone may be of equivalent potency but which, because of their different

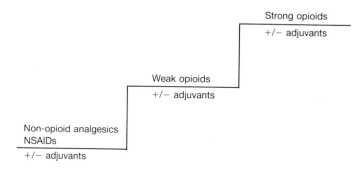

Fig. 6.1 The analgesic staircase.

modes of action in producing analgesia, may act synergistically to produce a combined effect greater than that of using either alone. For example, it may be pointless to change from codeine to dextropropoxyphene in a patient whose pain is inadequately controlled by the former, but if there is an element of inflammatory change or tissue destruction, then the addition of an anti-inflammatory drug such as ibuprofen may produce a far more acceptable degree of analgesia than its position on the analgesic staircase would suggest. Nevertheless, the staircase concept remains valid in that if a weak opioid drug is found to be ineffective, then logically one should proceed to a drug with a higher potency rather than changing to another weak opioid, whatever the stage of the disease or the life expectancy.

Non-opioid analgesics

The drugs usually considered in this group are aspirin, paracetamol, the non-steroidal anti-inflammatory drugs (NSAIDs), and possibly nefopam. Aspirin also has potent anti-inflammatory properties but may cause problems with sensitivity reactions, prolongation of bleeding time, and gastrointestinal erosion. The last of these problems is also a frequent complication of NSAIDs, which can occasionally have adverse effects on renal function. Nevertheless, all these drugs find a useful role in controlling cancer pain and may be all that is required. Aspirin or the NSAIDs can be profitably combined with opioid analgesics where their effect may be synergistic. As in all chronic pain situations, these drugs should be administered regularly and not *pro re nata*. The dose ranges are limited, and exceeding the recommended doses will not enhance analgesia.

Opioid analgesics

Opioid analgesics (the term is now both internationally preferred to 'opiates' or 'narcotics', although both are still frequently used) are perhaps one of the oldest, and yet most effective, medical treatments, although they are still frequently denied or prescribed in ineffective doses to those to whom they could bring relief. This is, no doubt, partly due to the increasing social abuse of this group of drugs, but it is also due to inappropriate prescribing in the past, as a result of inadequate understanding of the pharmacology of morphine. Recent advances in knowledge of the metabolism of morphine, and in technical achievements in pharmaceutical preparation, allow morphine to be more acceptable as a long-term means of managing chronic pain.

The weak opioids are usually regarded as the next step from non-opioid analgesics on the analgesic staircase, and are mainly represented by codeine, dihydrocodeine, and dextropropoxyphene. The role of these drugs is limited, as side-effects often predominate at low doses and, again, analgesia does not bear a useful relationship to dose. Codeine phosphate is prescribed in a dose of 30–60 mg 4-hourly. Dihydrocodeine is also used at a dose of 30–60 mg 4-hourly, although a slow release preparation is available to be given at 60 mg 12-hourly.

These compounds are often available in combination with aspirin or paracetamol and there may be a certain amount of synergy between them. It is suggested that the combination of 30 mg of codeine with 650 mg of aspirin exceeds the analgesic potency of 60 mg of codeine. Similarly, combination with 500 mg of paracetamol may have advantages. Dextropropoxyphene is commonly available in a dose of 32.5 mg with 325 mg of paracetamol (coproxamol) and this dose, or double this dose, every 4 hours may be sufficient in moderate pain. However, if pain is not readily controlled by these weak opioids or by their combination with non-opioid analgesics, then it is necessary to proceed without delay to the strong opioid analgesics. When patients present with severe uncontrolled pain, there is little point in delaying relief by a trial of the weak opioids, and treatment should start immediately with strong opioids, with or without the addition of other drugs. The concept of staircase analgesia should not preclude jumping the middle step.

One of the factors which has, perhaps, limited prescribing of morphine has been its apparent poor efficacy when administered orally. It is not uncommon for practitioners to prescribe parenteral morphine for short term relief and then, when trying to convert to oral medication, to find that oral morphine at the same dose does not relieve the pain. Oral morphine undergoes extensive first-pass hepatic metabolism so that changing from parenteral to oral administration of morphine at the same dose will result in greatly reduced plasma levels of morphine sulphate. This poor bioavailability of oral morphine led to the suggestion that a ratio of 1:5 should be used when converting parenteral to oral dose. However, it is now known that one of the main metabolites of morphine, morphine-6-gluconate, is a powerful analgesic and, as this product accumulates, the relative analgesic effect of oral morphine increases. This metabolite has a relatively long half-life, which has important implications when prescribing for patients who may be debilitated and have impaired renal and hepatic function. Some authorities believe that the bioavailability can approach 100 per cent (McQuay *et al.* 1983), but the most usually accepted ratio for conversion of parenteral to oral dose is 1:3 or even 1:2. These factors explain why oral morphine is a poor analgesic for relieving acute, short-term pain whilst for relieving chronic pain it reveals its true potency. Individual variation in morphine metabolism is sufficient to confirm the need to titrate the dose against the pain for each individual patient (Sawe *et al.* 1981).

Short-acting opioids such as pethidine, papaveretum, dextromoramide, or dipipanone are rarely useful for long-term administration in chronic pain. The resulting fluctuating blood levels provide short episodes of relief alternating with pain and symptoms of withdrawal. Maintainance of a constant blood level not only provides more acceptable relief from pain, but also helps to minimize undesirable side-effects. Drugs with rapid, but short-lived, action tend to promote drug-seeking behaviour (or pain-relief-seeking behaviour?) which is unpleasant for the patient and causes concern amongst those caring for the patient. Occasionally a short-acting drug, such as dextromoramide, may be helpful in providing enhanced analgesia at specific times, such as prior to painful nursing procedures, and its availability for such times may help to maintain the patient's confidence, providing they do not come to rely on it because their long-term analgesics are inadequate.

It is difficult to determine the role of partial agonists in palliative care.

Older drugs such as pentazocine have little place because of their short duration of action and side-effect profile. However, buprenorphine is occasionally valuable in the earlier stages. Its administration by the sublingual route is often acceptable to patients, and its length of action is beneficial. Apart from the problem of the nausea which frequently occurs during buprenorphine therapy, another problem also arises as the disease progresses: a ceiling effect of analgesia may be reached early on and increasing the dose thereafter merely increases the side-effects. However, a patient who has been maintained on buprenorphine can readily be converted to morphine if the need arises, despite the theoretical problem of mixing a partial agonist with a pure agonist, but it is obviously illogical to continue to administer the two drugs together for any length of time.

Although morphine and diamorphine have long been the main opioid analgesics used in managing cancer pain, one of their shortcomings has been a relatively short half-life. Chronic pain needs to be managed by timed administration of analgesics, each dose being administered before the effect of the previous one has waned. With morphine or diamorphine elixir, this means administering a dose every 3–4 hours. This problem has largely been overcome with the development of a sustained-release oral preparation which only requires 12-hourly administration but, until this was available, other drugs were available with a longer action for those in whom 4-hourly dosing was disliked, particularly where providing night-time analgesia without waking the patient after 4 hours was a problem. This same group of drugs is still useful for patients who are unable to tolerate morphine for one reason or another.

Sustained release oral morphine tablets (*MST Continus*) are now widely used for long term maintenance of opioid analgesia. Initial concerns about its bioavailability have been found to be groundless and it is as effective, milligram-for-milligram, as oral morphine sulphate solution (Walsh 1983).

Methadone has a long half-life and can be provided in liquid or tablet form. Some patients who are unusually sensitive to the nausea or depressant side-effects of morphine find methadone to be an acceptable alternative. It is usual to start with a dose of 5 mg 6-hourly and adjust the dose according to response. Unfortunately, the sedative effect of methadone frequently outlasts the analgesic effects and this can be cumulative, especially in the elderly. It is therefore advisable to monitor closely any patient starting methadone therapy and it is often possible, once a suitable analgesic dose has been determined for a particular patient, to decrease the frequency of administration to 8-hourly after a few days. This will reduce the level of sedation while maintaining adequate analgesia. Methadone is not popular with most hospices because of reported sedation. However, we have seen many ambulant patients, who had been intolerant of morphine, maintained satisfactorily as out-patients for long periods on methadone without untoward effect.

Phenazocine is useful in some patients who find morphine unacceptable. It can be administered in a starting dose of 5 mg 6-hourly and is well-absorbed sublingually, which may be an advantage in patients who find trouble in swallowing drugs. The unpleasant taste can be disguised by placing the tablet inside the hole of a mint.

Another method of providing longer-lasting analgesia, especially at night, is by the use of opioid suppositories. These drugs are well absorbed by the rectal route and, although delivery by this route does not necessarily prevent nausea, it may provide a reliable route of absorption if vomiting prevents the usual oral dose from being given. Morphine suppositories are readily available and the addition of a 10 mg suppository will often guarantee a good night's rest to a patient who is normally awoken by pain. Oxycodone can be obtained in this form through pharmacies on special request to the manufacturer and will provide an even longer period of analgesia; a 30 mg suppository will often produce 8 hours of analgesia. However, patients may develop proctitis with prolonged use of suppositories or find them unacceptable for other reasons. Diarrhoea renders suppositories less than effective and although suppositories have been usefully employed via a colostomy, use in this way may be limited.

Starting opioid therapy

Managing severe pain is more effective in patients who are in hospital than in those attending as outpatients, as the efficacy and side-effects of prescribing can be monitored frequently by the medical and nursing staff, and frequent adjustments made as required. However, in many cases, in-patient treatment is not possible but frequent visits, if combined with good communication with the GP and district nurse, can be satisfactory.

Many units start their in-patients on 4-hourly morphine elixir, often 5 mg/ml initially, and increase the dose steadily until adequate analgesia is obtained or the dose becomes unacceptable. If desired it is then possible to convert from a 4-hourly dose of morphine sulphate elixir or diamorphine to 12-hourly sustained-release morphine. The starting dose will depend on many factors, such as renal and hepatic function and, in some debilitated patients, reduced level of plasma albumin with consequent reduction in binding of some drugs, so altering the required dose. However, for opioid-naive patients, 5–10 mg 4-hourly is the usual starting dose. Patients who have already been taking opioids but whose pain was inadequately controlled should have the previous dose converted to an equivalent dose of morphine, and this dose then increased by 50 per cent. In severely debilitated patients, the dose may have to be increased more gradually (Regnard and Davies 1986). Once the patient is taking 4-hourly morphine, the dose can be titrated against the pain by increasing it once or twice a day in increments of 30–50 per cent. There is no standard dose, and the required amount covers a huge range. Some patients will require doses of several hundred milligrams, although the majority will be controlled on less than 200 mg per day. However, although large doses are sometimes required, it should be remembered that not all pains are entirely morphine-sensitive, and alternative or adjuvant drugs or therapies may be necessary if pain control is difficult.

When managing pain in the out-patient, it may not be easy to titrate 4-hourly morphine against the patient's symptoms unless there is good home nursing backup with freedom to vary medication according to the response. It should not be assumed that such freedom of nursing discretion will be given unless this has become standard practice in the district. If the hospital doctor has to assume responsibility for determining the dose of opioid analgesics, it may be helpful to use slow-release morphine from the beginning. However, this is only possible when the hospital doctor is able to regularly and frequently assess the response, and is readily available. If this cannot be guaranteed, or the response in a particular patient is likely to be particularly unpredictable, then it may be better to admit the patient for a short period of stabilization.

A 4-hourly dose of morphine may not provide adequate cover throughout the night. This can either be overcome by using a longer-acting preparation, using suppositories, or simply doubling the bedtime dose of morphine sulphate elixir, and omitting the middle of the night dose (Twycross 1982).

Morphine elixir is convertible milligram-for-milligram to slow-release (SR) morphine, and it is a simple matter to calculate the total daily dose of elixir and then divide into two equal 12-hourly doses. Occasionally physicians are tempted to increase the frequency of administration of SR morphine, when the level or duration of pain relief is inadequate. However, the *Continus* formulation of the usual form of SR morphine (MST) is designed to maintain a reasonably constant level over 12 hours, and it is advisable to increase the 12-hourly dose rather than its frequency. Only rarely is it necessary to use an 8-hourly dose regimen. It is important to bear in mind that much of the analgesia provided by morphine administration in chronic pain is related to the activity of longer-acting active metabolites.

Alternatively, if there is an urgent need to rapidly determine the effective dose of morphine to control pain, then incremental doses of morphine can be administered intravenously until satisfactory analgesia is obtained, and the oral dose then started at a dose of three times more than the intravenously determined dose.

When starting opioid therapy, nausea is a common side effect and it may be advisable to prescribe an anti-emetic routinely. However, tolerance to the emetic effects of opioids usually develops over a few weeks and it may then be possible to discontinue routine anti-emesis. The commonly-used anti-emetics such as prochlorperazine, metoclopramide, and domperidone are often sufficient and, if necessary, they can be given for a short period by suppository. However, if nausea persists as a result of opioid therapy, then haloperidol in a dose of 0.5–5.0 mg twice daily may be more effective.

Sedation may accompany the start of opioid therapy but the patient can be reassured that this generally subsides over a few days. If it persists to an unacceptable level, then it may be necessary to reduce the dose of analgesic to a better tolerated level and then increase it again gradually as tolerance to sedation develops.

Constipation is commonly a complication of opioid therapy and routine use of a bowel stimulant and faecal softener is recommended. A combination of a stimulant laxative such as danthron with a softener such as poloxamer or docusate is usually effective. However, it is important not to administer such preparations if there is a possibility of intestinal obstruction or the patient complains of colicky abdominal pains. It is occasionally necessary to resort to enemas or manual disimpaction in severe cases.

Although there is no correct dose range for morphine or diamorphine when used in chronic pain, other than that which satisfactorily controls symptoms without unacceptable side-effects, if very large doses are necessary, consideration should be given to whether this is a type of pain which is unresponsive to opioid analgesics, or whether the addition of adjuvant drugs may enable a reduction in opioid dose. The concept of opioid-insensitive pain is a clouded issue and much more investigation is needed in this area. However, common experience indicates that pain arising from invasion of bone or damage to nervous tissue by tumour is, at best, only partially responsive to opioid analgesics and is best managed by the addition or substitution of other drugs (see p. 62). Acute pain superimposed on chronic pain, such as arises on movement of a pathological fracture (sometimes termed 'incident pain'), is unlikely to be supressed with opioids while the patient remains rousable, and other methods must be employed.

The debate about the relative values of morphine and diamorphine in palliative care will probably continue, but controlled trials have shown little advantage of the latter over the former (Twycross 1977). Slight differences in patient acceptability, such as reduced nausea, are sometimes suggested but any differences are probably minimal. There is a general desire in the hospices to promote education in the use of oral morphine instead of diamorphine, as diamorphine is unavailable for medical use in large areas of the world. However, diamorphine is valuable for parenteral administration owing to its vastly greater aqueous solubility than that of morphine, so that large doses can be administered in small volumes, which is especially useful when subcutaneous infusion is employed.

Morphine elixirs are frequently flavoured to increase palatability but, unfortunately, pharmacies commonly supply morphine elixir in chloroform water as a preservative, which is responsible for much of the unpleasantness. It has been suggested that the addition of chloroform water is not absolutely necessary and that the shelf life of morphine elixir is probably longer than often stated.

The addition of other drugs to morphine elixir is nowadays regarded as undesirable. Traditionally a cocktail (Brompton Cocktail) was made, containing such additions as chlorpromazine, cocaine, and the patient's favourite spirit (often gin or brandy). The routine addition of chlorpromazine as an anti-emetic (or, as it was previously thought, to enhance the analgesic effect of the opioid) results in excessive sedation and limitation in flexibility of dose of the mixture. Cocaine was added for a variety of well-meant reasons, particularly as a euphoriant. Unfortunately it may act as a dysphoriant. The alcohol, whilst for some improving palatability, again reduces flexibility in dose and may have unpleasant side-effects. If alcohol is administered in palliative care, and there is no doubt that it may play a valuable role as an appetite stimulant, mild relaxant, and social catalyst, then the more conventional form is more controllable and enjoyable. In many hospices the pre-lunch sherry or the night cap is a popular and effective addition to overall patient care.

Table 6.2 Approximate equivalents to morphine sulphate of other opioid drugs

| Drug | Approximate equivalents to morphine sulphate 10 mg 4-hourly | |
	Dose	Administration schedule
diamorphine	7.5 mg	4-hourly
CR morphine (MST)	30 mg	12-hourly
oxycodone suppository	30 mg	8-hourly
buprenorphine	0.2 mg	6-hourly
dextromoramide	5 mg	3-hourly
phenazocine	5 mg	6–8-hourly
levorphanol	1.5 mg	6–8-hourly
methadone	10 mg	6–8-hourly

There are times when the oral route for the administration of analgesics becomes unsuitable (Fig. 6.2). The commonest reason for this is inability of patients to retain oral medication due to nausea and vomiting, or when absorption is ineffective due to gastrointestinal obstruction or stasis. The development of the portable syringe driver has enabled analgesics or other drugs to be delivered parenterally to patients without the discomfort or inconvenience of repeated intramuscular injections, while at the same time

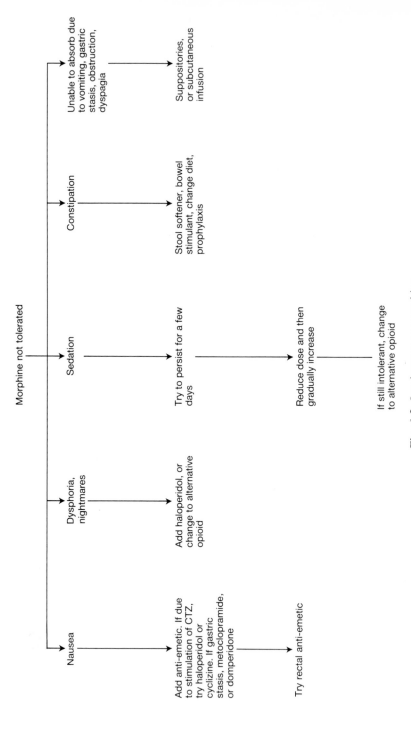

Fig. 6.2 Intolerance to morphine.

providing the advantage of a steady plasma level of the drug (Oliver 1988; Oliver 1985*a*). This latter feature may be one of the reasons why occasionally it is possible to control pain effectively with administration of analgesic via a syringe driver, when other routes of administration have been ineffective or have been impossible because of intolerable side-effects. In a few instances, it is therefore helpful to introduce drug delivery by syringe driver even in patients who are capable of taking oral medication, but in whom control is poor. In these circumstances, if good pain control can be achieved via the continual subcutaneous route, it is usually possible to successfully reintroduce analgesia via the oral route. The syringe driver also finds a useful role in continuing medication during the last days of life if the patient is no longer able or conscious enough to co-operate with oral medication. The continuation of medication to control distressing symptoms in this way can be a great comfort to relatives.

There is an increasing number of syringe drivers on the market, but the essential requirements are that they should be lightweight and easily portable for the patient, simple to use for the attendants, and reliable. The portable pumps can be comfortably worn in a shoulder holster by the patient. There should be some indication on the pump, by audiovisual warning, when the syringe is empty or when the battery is low.

The pumps most commonly used take a disposable 10 ml syringe which is filled with the patient's 24-hour requirement of analgesic. If all the drugs are contained in a standard 10 ml volume the setting of the pump is simplified and there is less chance of errors being made by medical or nursing staff. When the dose is changed it is better to start a new syringe with a different concentration of drug than to alter the rate of infusion. It is important that the infusion is kept to this minimum volume, as subcutaneous administration of larger volumes over 24 hours will increase the incidence of local skin problems. For this reason diamorphine is the preferred analgesic owing to its tremendous solubility compared with morphine.

Once the syringe has been filled, it is loaded into the driver and connected to a long-tailed, winged, infusion needle, and the tubing is primed. The needle is inserted subcutaneously, commonly on the upper chest wall or sometimes over the flank of the abdominal wall. A loading dose should be given initially, and this should be equivalent to the dose that would normally be given over a 4-hour period by alternative routes. The priming volume and the loading dose necessary when starting an infusion should be accounted for when loading the first syringe, so that there is still 10 ml in the syringe to be delivered over 24 hours. The needle can be secured to the skin with a non-irritant adhesive dressing and may comfortably remain in place for considerable periods.

In most types of pump, a type PP3 battery will deliver about 50 full syringes, and therefore should last for about 6–7 weeks. On the Graseby pumps, a flashing light which indicates correct function ceases then the battery is low, but the pump should continue to function for approximately 24 hours after this. On some models there is a boost button which can be used to deliver a bolus.

Provided that good skin cleansing is performed prior to insertion, the needle site may be maintained for long periods, often for many weeks. Induration may develop, but it is only necessary to change the site if inflammation is obvious. There are many reports of subcutaneous infusions being used for periods of 6 months or more without problems. Local reactions at the site of infusion are more common when other drugs are added to the diamorphine, and these tend to be chemically-induced inflammatory reactions.

Several other drugs may be added to the diamorphine for administration by subcutaneous infusion. Cyclizine, haloperidol, methotrimeprazine, and hyoscine are all suitable. Chlorpromazine, prochlorperazine, and diazepam are unsuitable as, being highly irritant, they frequently produce local inflammatory reactions. Occasionally, when used in high concentrations, cylizine and diamorphine may produce a crystalline precipitation. Whichever drugs are used, it is wise to change the syringe daily, especially when two drugs are mixed in the syringe.

Although syringe drivers have proved to be a most valuable aid in the treatment of pain in malignant disease, they are occasionally used unnecessarily, especially when they are in plentiful supply! Most patients can be managed on oral medication, and there must be a positive indication for changing to a parenteral route.

Opioid-unresponsive pain

Not all pain responds to opioid analgesia, although most pain in malignant disease is a mixture of different types, and some components may respond leaving an unrelieved component which requires the addition of other drugs or other non-pharmacological intervention. The types of pain where opioids are generally helpful tend to be those where there is an active nociceptive process continuing. This may result from invasion of normal tissue, or the distension or compression of viscera. Opioids are rarely adequate to control neurogenic pain or pain resulting from invasion of bone. They tend to be more useful for managing the aching, boring, pressing types of pain, and less helpful for burning, shooting pains or those sharp pains resulting from movement of pathological skeletal tissues.

Non-steroidal anti-inflammatory drugs (NSAIDs)

These drugs have a potent inhibitory effect on prostaglandin synthesis and this makes them a useful adjunct in the management of pain involving an inflammatory process. They can be used as mild analgesics alone, or used together with opioids to produce a greater combined effect. They are especially useful in managing pain from bone metastases. Aspirin is cheap and effective at standard anti-inflammatory doses but may be poorly tolerated by the gastrointestinal system, especially in already debilitated patients. Gastrointestinal bleeding is a hazard with all NSAIDs, but may be slightly less of a hazard with the salicylate-based diflunisal. This is given at a dose of 500 mg 12-hourly, which fits conveniently with the dose regimen for slow-release morphine. Alternatively, drugs such as flurbiprofen 100 mg 12-hourly, or piroxicam 20 mg once daily may be preferred. There is no ideal NSAID, and individual preferences of the patient and familiarity with the doctor are probably important determinants of which preparation is chosen. Diclofenac and piroxicam are available as suppositories, when the oral route presents problems.

NSAIDs find a role in the management of localized cutaneous inflammation and pruritis in some malignant conditions, and also in reducing the excessive sweating of which some patients with malignant disease complain, especially at night.

Corticosteroids

Corticosteroids are valuable drugs for symptom control in malignant disease, with a variety of indications. Their anti-inflammatory action may help to control pain where there is extensive tissue destruction and invasion. This is particularly true in pain resulting from nerve compression or destruction, as this pain is often poorly responsive to opioids. The pain from raised intracranial pressure frequently responds well to high-dose steroid therapy, and pain from distension of the liver capsule is a definite indication for steroid therapy. Apart from these specific indications, low doses of steroid (e.g. dexamethasone 4 mg daily) are often useful in improving appetite and general feeling of well-being. The loss of appetite and weight is a common cause of distress in these patients.

Dexamethasone is the steroid of choice, as it can be conveniently administered in high dose and has fewer undesirable side-effects than some other preparations. For managing pain from nerve compression or raised intracranial pressure, the patient should initially be prescribed 12–24 mg of dexamethasone daily in divided doses. The last dose of the day should not be given after 6 p.m. as this may result in insomnia. If useful pain control is achieved, the dose can usually be reduced after a week, and a maintainance dose established. Side-effects do occur, but these are commonly in the form of weight increase and fluid retention. Candida may be a problem which can be treated effectively, but the more serious effects of gastric erosion are fortunately uncommon enough (perhaps 5 per cent) to justify the risk where the benefits of such treatment are substantial. The concurrent administration of corticosteroids and NSAIDs is likewise occasionally justified.

Antidepressants

Antidepressant drugs are used in three main areas in the management of malignant disease. As with any chronic disease state, especially with a chronic pain syndrome, depression is a common symptom and its appropriate treatment may allow more successful management of other symptoms. Some patients experience symptoms of depression as a result of opioid therapy. There is also frequently a degree of anxiety present, and these symptoms along with pain and other uncomfortable problems may deny the patient an adequate night's rest. There are often, therefore, advantages in choosing an antidepressant that also has anxiolytic and sedative properties which, when the total dose is administered as a single evening dose, may ease several symptoms. Amitriptyline tends to be the standard by which other antidepressants are compared, and this is certainly a valuable agent. Oversedation may be troublesome, and it is wise to start with a low dose initially and increase over a few days. In frail or elderly patients the starting dose can be as low as 10 mg at night, and if necessary increased in stages to 75 mg. In these circumstances it is uncommon to require a higher dose than this. If daytime sedation continues to be a problem, dothiepin may be more suitable, and is usually well-tolerated at a starting dose of 25 mg at night. The tetracyclic drug, mianserin, given as a dose of 30 mg at night may suit patients who cannot tolerate the anticholinergic effects of the tricyclics, especially if dry mouth becomes intolerable or if urinary retention is a problem. The antidepressants are often more suitable a sedative in the restless, depressed insomniac patient with malignant disease than are the benzodiazepine hypnotics. Imipramine may be a better choice if no sedative effect is desired, and depression appears to be the major problem.

Apart from these more obvious uses of the antidepressant drugs, they frequently have a valuable role to play in the management of pain syndromes. Although they are not classed as analgesics, the tricyclic antidepressants have a central analgesic effect in certain types of pain. They are particularly useful in dealing with pain of a neurogenic origin, especially when the pain is of a burning dysaesthetic nature. Such pain may follow nerve damage resulting in deafferentation of a region, or when nerves are partially damaged, resulting in altered sensation such as allodynia or hyperpathia. Some pain appears to be accompanied by altered reactivity in the sympathetic nerve supply and this may also be reduced by tricyclic therapy. This is a property which is either not present in the antidepressants of other groups, or is certainly only a weak effect. Unfortunately, perhaps, the drug which is most effective in this area is amitriptyline, which is also the drug which is sometimes less well-tolerated than many others. However, if the drug is introduced in low dosage and increased gradually, tolerance can usually be achieved. If amitriptyline is intolerable to the patient, then a second choice would be dothiepin, which appears to be only slightly less effective. If used to treat pain, tricyclic anti-depressants may be effective in low doses, but in some resistant cases it is necessary to use high doses (e.g., 150 mg amitriptyline) to gain a useful effect.

Tricyclic antidepressants may be useful alone when dealing with burning or dysaesthetic pain, but they are often combined with opioid therapy in treating pains which are not well-controlled by opioids, usually because of the element of nerve damage.

Phenothiazines

Phenothiazines do not have a large place in managing pain. Traditionally chlorpromazine was combined with diamorphine (in Brompton Cocktail) and its use was believed to enhance the analgesic effect of the opioid. This is not now generally held to be true, but it may be valuable as an anti-emetic or in the treatment of intractable hiccoughs. Sedation and postural hypotension are unfortunately common accompaniments to the use of this drug.

Of more interest is methotrimeprazine. Again it has been suggested that this drug has analgesic properties, and this may or may not be true. The main indications for its use are terminal agitation and restlessness, and severe vomiting. The agitation shown by some patients towards the end of their lives is distressing to both patient and carers and methotrimeprazine is more successful in controlling this symptom than chlorpromazine or diazepam. When given with diamorphine it appears to enhance the analgesia and is a potent anti-emetic. The oral form is no longer available but the injectable drug is physically compatible with diamorphine and can be easily administered with this in the same syringe for continuous subcutaneous infusion. The dose range is 25–100 mg per 24 hours, with the lower dose being adequate in the majority of cases. The common side-effect is sedation, but where this is acceptable it is a valuable drug for controlling particularly distressing symptoms (Oliver 1985b).

Anticonvulsants

Anticonvulsant drugs play a major part in the management of pain which is believed to have a component due to nerve damage. This may be as a result of trauma or invasion of peripheral nerves, or in the central nervous system, where the overall effect of the damage to nervous tissue results in pain of a sudden lancinating nature. This is the pain which is commonly described as 'shooting' and the patient may liken it to electric shocks. Occasionally other pains of a neurogenic nature may respond to anticonvulsant therapy, if the pain is of an intermittent sharp nature, but not the type of pain resulting from pathological fractures (or incipient fracture).

The classic drug described for use in lancinating pain is carbamazepine, as has been well recognized in its use to control trigeminal neuralgia. This may be prescribed when pain from malignant disease has a lancinating component, starting at 100 mg 8-hourly if tolerated, and increased to 200 mg 8-hourly if necessary. However, some patients are unable to tolerate this and it may be necessary to start at 100 mg at night and build the dose up more gradually. Sodium valproate is becoming a popular agent to use in managing this type of pain, and we often find this drug to be not only more effective, but better-tolerated in many cases. Again the starting dose may need to be as low as 100 mg once a day, but it is usual to start at 200 mg 8-hourly, and if necessary this can be increased incrementally to 600 mg 8-hourly. It is often necessary to use the higher dose ranges of these drugs to gain adequate control of lancinating pain. Further alternatives are phenytoin, and clonazepam, although the latter is often too sedative in elderly or debilitated patients.

All of the anticonvulsant drugs may produce troublesome side-effects, most commonly gastrointestinal upsets and sedation. Sodium valproate, although perhaps the most useful of this group, occasionally can produce blood dyscrasias or disorders of liver function, and so the blood count and liver function tests should be monitored. The occasional side-effect of alopecia produced by sodium valproate therapy is generally reversible when the drug is stopped.

The anticonvulsant drugs are frequently required in conjunction with other 'coanalgesics' such as antidepressants, and may be prescribed concurrently with opioid analgesics if necessary.

Other painful symptoms

The pain from bone metastases has already been mentioned as being particularly difficult to control and various other measures may be required. Pain from invasive tumour is a complex situation with a wide range of chemical mediators being produced at the site of osteolysis, and it is not entirely clear exactly what is the mechanism of pain generation. Single-treatment radiotherapy, either of the isolated lesions or of a wide skeletal field, may produce rapid relief of pain even before there is any evidence of resolution of the lesions or healing. Radiotherapy is always worth considering in such situations. If a pathological fracture has occurred or is imminent, then surgical stabilization should be considered, especially in the long bones or where vertebral collapse is resulting in pain. Although one of the major considerations in dealing with pain in late-stage cancer should be to avoid unnecessary investigations and treatment, the timely involvement of an orthopaedic surgeon may enable the patient to regain mobility and obtain pain relief and, if this can be achieved, many patients with bone secondaries can survive for considerable periods with acceptable quality of life.

There is anecdotal evidence that some patients may obtain prolonged relief from the pain of bone metastases when single deposits are directly injected with steroid. This may be technically impossible, and there is no controlled trial that demonstrates its effectiveness but, in cases where a single bone secondary is accessible to direct injection, this technique deserves further consideration.

Pain from widespread bone metastases may respond to either treatment with exogenous endocrine hormones, or by manipulating the patient's own endocrine system. The introduction of salmon calcitonin into the management of pain from malignant disease has been a controversial topic. There is no doubt that in patients where hypercalcaemia accompanying osteolysis causes the symptoms of thirst, polyuria, drowsiness, and nausea, then treatment with calcitonin along with rehydration will tend to relieve symptoms. However, salmon calcitonin does appear to have some specific analgesic properties when used for treating bone metastases (Hindley *et al.* 1982). Patients should be given 400 units by intramuscular injection 12-hourly for 48 hours. They usually require regular treatment with anti-emetics during this period. Maintenance can then be in the form of a weekly dose of 400 units administered by the district nurse. This treatment is expensive, but sometimes appears to be extremely effective in resistant cases. Unfortunately, the controlled trials reported so far do not offer completely conclusive evidence of the efficacy of calcitonin.

A similar role has been reported for biphosphonates (Morton *et al.* 1988), and early reports suggest an analgesic effect and sclerosis of lesions, even when plasma calcium and phosphate concentrations remained unaltered.

Widespread pain from malignant disease may be managed by a partial destruction of the pituitary. It was originally noted that patients with hormone-dependent breast tumours undergoing pituitary ablative procedures often obtained effective pain relief. At first this was thought to be due to the effect of tumour regression, but it was noticed that the pain relief was often of very early onset before any appreciable tumour regression could have occurred, and also that tumours which were not hormone dependent could also be effectively treated in this way to provide analgesia. Many mechanisms for this analgesic effect have been proposed. It was suggested that the alcohol injected into the pituitary had its effect by spreading to the hypothalamus, but pituitary destruction by cryotherapy or electrical lesions are equally effective. It is possible that changes in the activity of the endogenous opioid systems are involved, but much of the evidence is still inconclusive. Although pituitary lesioning is often an effective means of relieving widespread pain from bone secondaries, the technique is not

without considerable morbidity and is less popular now than it was a few
years ago.

Pain from intestinal obstruction

Intestinal obstruction is a common event in the development of intra-
abdominal tumours, either when they originate within the gut, or when
obstruction is due to pressure from without. Traditional treatment is either
by surgical intervention or conservative management by a 'drip and suck'
regime. The vomiting and abdominal pain that normally accompany obstruction are particularly distressing and the urge to intervene may be overwhelming. Obstructions due to simple adhesions in a patient who may otherwise
have reasonable life expectancy may well be best dealt with by timely surgical
intervention. However, there are many patients who are unwilling to undergo
further surgery or in whom, because of the advanced nature of their disease
or the inoperable nature of their obstruction, it is considered best to adopt a
conservative management plan. The discomfort of a nasogastric tube and
intravenous infusion in the last days of life may often be avoided without
undue problems, provided that nausea and pain are controlled (Baines *et al.*
1985). The patient with obstruction of the large, or even small bowel may be
allowed small amounts of oral fluids to provide comfort, and sufficient
amounts of fluid may be absorbed by the upper gastrointestinal system to
allow adequate hydration for a period. Nausea should be controlled with
anti-emetics and, if this symptom is controlled, most patients will tolerate the
occasional vomit without too much distress. If obstruction occurs high in the
gut, then gastric distension may result in unacceptable vomiting and dehydra-
tion, and a nasogastric tube will be necessary.

The pain of intestinal obstruction may be satisfactorily reduced by
reducing excessive bowel activity. This means avoiding laxatives which
stimulate the bowel (such as senna or danthron) and stopping osmotic laxatives such as lactulose which may aggravate gut distension. Anti-emetics are
generally required, but metoclopramide and domperidone should be
stopped, as these stimulate the upper gut to act against the obstruction at a
lower level. Cyclizine at 50 mg 8-hourly (subcutaneously or per rectum) is
the first choice of anti-emetic in this situation. If this is ineffective,
haloperidol can be added in a dose of 5 mg at night.

If intestinal colicky pain persists, it may be useful to prescribe hyoscine.
This can be given as the hydrobromide, either 0.3 mg sublingually 8-hourly,
or 0.8 mg subcutaneously 8-hourly. Hyoscine butylbromide (*Buscopan*) may
be equally effective in a dose of 10–20 mg 6-hourly subcutaneously whilst
producing less dry mouth than with the hydrobromide.

Constipation as a cause of obstruction and abdominal pain should always
be considered, especially in patients who have been treated with opioids, and

the use of a stool softener (docusate 60–180 mg) plus a mild stimulant (danthron 50–150 mg) is valuable at an early stage, provided that there is no other cause for obstruction. These drugs are available in a combined preparation. It is only occasionally necessary to resort to enemas and manual evacuations as a temporary measure.

Other symptoms

A variety of symptoms may present in the patient with malignant disease, which although not necessarily painful, can be extremely distressing to the patient, and it therefore seems appropriate to briefly mention some of them in this text. Many excellent works are available on symptom control in malignant disease, and the reader is referred to these for further information (Regnard and Davies 1986; Twycross and Lack 1983, 1986; Baines 1978). Nausea and vomiting are common problems, and the nausea experienced by some patients may be at least as, if not more, distressing than the pain which a patient suffers. Apart from the subjective effects of these symptoms, there is the added problem of difficulties with oral intake of food and fluids, and especially a problem with oral medication.

Nausea may result from a variety of causes, and the management may depend on the causes. Intestinal obstruction has been referred to above, and although the eventual management may or may not be surgical, it is usually necessary to provide some relief from nausea, especially when the obstruction is in the upper gut, even if vomiting cannot be prevented. As mentioned before, anti-emetics which increase gut motility should be avoided, and this means metoclopramide and domperidone. Cyclyzine is probably the drug of choice here, although it may be necessary to add haloperidol or hyoscine.

Much of the nausea in malignant disease results from stimulation of the chemoceptor trigger zone (CTZ), either by drugs such as opioids (or chemotherapeutic agents, although this is not usually a factor late in the disease) or as a result of the disease process. The nature of the latter is not understood, but it seems that the destruction of tissue may release chemical factors which stimulate the CTZ. The phenothiazines are often sufficient to control nausea, and the usual doses of prochlorperazine or perphenazine may be used. If the oral route is precluded by vomiting or gastric stasis, prochlorperazine may be given as a suppository. If there is known to be no obstruction, then domperidone may be effective given as a suppository. If these drugs are ineffective, haloperidol is perhaps the most useful drug. It can be given orally or by subcutaneous injection and doses as small as 0.5 mg daily may be sufficient. More intractable cases may require as much as 5 mg once or twice a day. Cyclizine is another alternative. Parenteral routes are often necessary until vomiting has been controlled, or gastric motility re-established, but in many cases it will be possible to return to and maintain the patient on oral medication. Where the oral route is excluded for any length of time, and a suppository form is either not available or is unsuitable, then a continuous subcutaneous infusion should be considered. Haloperidol and cyclizine are particularly suitable for delivery by this route, and they can be combined with opioid analgesics in this way. If nausea is resistant to these drugs, and especially if agitation is a problem in the later stages of disease, methotrimeprazine is a potent anti-emetic and sedative and is ideally given via the syringe driver.

Insomnia in malignant disease may result from any of the other manifestations of the disease, and if relief from pain and other discomforts are achieved, then sleep may follow. Nevertheless, some patients are still unable to obtain a good night's sleep, and other measures may be required. Often this involves personal contact with reassurance, company, and perhaps spiritual counselling, but pharmacological help may be required. A benzodiazepine hypnotic such as temazepam may be all that is required, but patients have often been on these for some time already, and no longer find them effective. If night fear, disorientation, or nightmares are troublesome, then a sedative drug like haloperidol is often helpful at night or, especially when anxiety and depression are prominent, a sedating antidepressant may be the best choice. Amitriptiline in a dose of 10–100 mg at night or dothiepin 25–100 mg are good choices.

Sweating may occur in malignant disease, especially at night, the patient soaking the bed linen, and this can add to the other discomforts. If it is necessary to treat this symptom, the addition of a small dose of NSAID may be helpful.

Diarrhoea, if it is to be treated symptomatically, usually responds to loperamide 2–4 mg, or the more traditional codeine phosphate. Any infectious or inflammatory conditions causing the diarrhoea may need to be treated concurrently. Rectal inflammation following radiotherapy may be helped by the addition of steroid therapy. Painful tenesmus can sometimes be relieved by low doses of chlorpromazine.

Cough and dyspnoea may require opioid therapy for control. Methadone is the traditional potent antitussive, but most powerful opioids can be effective. Treating dyspnoea with opioids is to provide subjective relief for the patient, and it is rare to depress respiration to a dangerous level when giving a dose sufficient to provide relief. Dyspnoea resulting from pleural effusion can be relieved with drainage but, if the symptom is due to lymphangitis carcinomatosis, high-dose dexamethasone may be effective.

Pruritus is usually managed with the antihistamine type of drugs. However, where pruritus is associated with cutaneous inflammation, the use of a NSAID is recommended.

Lymphoedema following occlusion of the lymphatic supply to a limb may result from direct invasion by tumour, fibrosis, or surgical interference. The discomfort experienced with this condition may be severe enough to be described as pain. Treatment is by elevation of the limb, frequently-repeated exercises of the limb muscles, and intermittent compression using one of the commercially available inflatable devices which help to pump fluid from the limb. (Information about these appliances can be obtained from The Mastectomy Association of Great Britain, or British Association of Cancer United Patients [BACUP].

Indications for, and common side-effects of, the various drug groups discussed are summarized in Table 6.3.

Table 6.3 Indications for, and side-effects of, the various drug groups

Drug	Indications	Common side-effects
NSAIDs	tissue destruction, inflammation, pruritis, sweating, bone erosion	gastrointestinal and renal problems
steroids	nerve compression, raised intracranial pressure, anorexia, weight loss, pelvic invasion, cord compression, bone pain, liver distension	increased weight, fluid retention, insomnia
muscle relaxants	muscle spasm	sedation, especially with diazepam
antidepressants	dysaesthesia, burning, night sedation, depression	sedation, dry mouth, urinary retention
anticonvulsants	lancinating, shooting pains	gastrointestinal upset, blood dyscrasias
sedative	agitation, restlessness	somnolence, hypotension
bone metabolic drugs	bone pain	nausea

Invasive procedures

The term 'invasive procedures' generally refers to nerve blocks, but there are also other procedures such as pituitary ablation and cordotomy which should be included.

The use of invasive procedures has an important place in pain relief, especially from the historical point of view. Phenol injections into the subarachnoid space and other parts of the central and peripheral nervous system were for many years amongst the few effective treatments available for intractable pain, and the technical challenge drew many of the early pioneers to this field. In many countries this situation prevails, but in regions where opioid analgesics are available for medical use, the growing awareness of the efficacy of these drugs and the more realistic appraisal of their dangers has led to the development of effective pain relief techniques which avoid invasive and destructive procedures. Coupled with the better understanding of the use of 'non-analgesic' drugs in pain management and pharmaceutical advances in preparation and delivery of drugs, these changes mean that the place of invasive procedures has diminished.

It is also important to appreciate that destructive procedures for pain relief have many disadvantages. Generally they are irreversible, and tend to be all-or-nothing techniques. Destruction of part of the pain perception system often results in unwanted destruction of other parts of the nervous system, and loss of motor power or continence may be an unacceptable price to pay for analgesia. If total anaesthesia is achieved with a nerve block the area of numbness may prove to be unexpectedly unpleasant for the patient and, with time, the central nervous system responds to deafferentation by changes in the nociceptive systems which may result in the development of unpleasant dysaesthetic pain, perhaps more distressing than the original pain. Finally, invasive procedures remove from the patient any sense of involvement in managing their pain, which may be an important factor in some patients. Thus, although nerve blocks do have a role in the management of pain in malignant disease, the indications for such treatment should be carefully considered in each patient, and the limitations of these measures appreciated.

A destructive nerve block may be appropriate when a pain is localized to one particular area with a well-defined nerve supply, which is relatively easily accessible, and where a selective denervation is possible without unacceptable damage to motor systems. A lesion which can selectively destroy nociception whilst leaving touch sensation relatively intact has obvious advantages. A destructive lesion should only be considered after other methods of management have been considered to be inappropriate or ineffective and, because of the late effects of deafferentation, a major destructive lesion, as in a cordotomy, may only be suitable in patients who have a limited life expectancy. Some destructive procedures which are known to have a specific and more predictable benefit such as coeliac plexus block for pain from pancreatic tumours may be considered almost as a first line treatment, but generally it is best to try more conservative measures first.

Some invasive procedures are relatively easy to learn and perform, but nearly all destructive procedures should be performed with the aid of an image intensifier. This helps to make a potentially damaging procedure into a more controlled technique, and where facilities are available it is rarely justified to attempt the more complex procedures without radiological control. The exceptions may be a few simple peripheral blocks, or occasionally a subarachnoid block in a patient in the final stages of a disease in which pain is uncontrolled, and mobility and continence are already severely compromised. All nerve blocks involving more than small-volume injections of peripheral nerves should be accompanied by the usual monitoring and resuscitation facilities, except in certain extreme circumstances. Some books produce vast lists of nerve blocks, but for the majority of patients with pain from malignant disease there are a relatively small number of blocks which will be regularly needed. Some of the more complex procedures such as cordotomy are not required sufficiently frequently to enable a practitioner working in a small unit to maintain adequate expertise, and it may be that these lesions are best performed in regional referral centres. However, this is probably a controversial point as most doctors have their favourite techniques which they are reluctant to cede to others. There is certainly no substitute for 'hands on' experience for invasive procedures, and most techniques are best learnt by personal instruction and practice. Also, there are many texts which describe in great detail the practical aspects of blocks. It therefore seems inappropriate for this work to provide more than an outline of some of the procedures, but it is important to discuss the indications and effects of invasive procedures.

Coeliac plexus block

The coeliac plexus is a loose collection of ganglia which lies anterior to the body of the first lumbar, and to some extent the twelfth thoracic, vertebrae. The aorta lies behind the plexus, and slightly anteriorly and to the right lies the inferior vena cava and right renal vessels. The pancreas and left renal vessels lie to the left. The plexus contains visceral afferent and visceral efferent sympathetic fibres although parasympathetic fibres pass through the structure. The greater, lesser, and least splanchnic nerves feed into the coeliac ganglia. The plexus is surrounded by loose connective tissue allowing injected solutions to spread readily throughout the structure. However, adequate spread of a neurolytic solution may be prevented by tumour mass or post-operative adhesions, and for this reason the results are often better when the injection is performed at an early stage of the disease. The coeliac plexus contains afferent fibres from the upper abdominal organs and destruction of these tracts will theoretically relieve pain from malignant disease in these organs. It is particularly useful for tumours involving the pancreas, liver, and stomach, and blockade of this plexus can be one of the most effective pain-relieving procedures. Even if complete analgesia is not obtained, it is usually possible to reduce pain to a level that can be adequately controlled with opioids. It is one of the few destructive procedures which is indicated soon after a lesion in these organs has been confirmed, and pain is becoming a major symptom. Some surgeons are prepared to inject the plexus intraoperatively when possible, but the technique is commonly performed as a percutaneous procedure. There is little justification for not using radio-logical control in what can be a potentially hazardous procedure. Some have advocated the use of computerized tomography (CT) scanning to confirm needle position, but this is rarely feasible or necessary, and a portable image intensifier in the theatre is quite adequate. The procedure is potentially painful, and the prone position may be uncomfortable for the patient with an abdominal malignancy. We therefore prefer to perform this procedure under a general anaesthetic if facilities are available, and if the patient can tolerate an anaesthestic. It is perfectly possible to inject the coelic plexus under local anaesthesia, but the sedation that may be required may be more of a problem to the debilitated patient than a light, modern, general anaesthetic.

The patient is placed prone on a radiolucent operating table, preferably with a pillow placed under the patient in order to slightly flex the spine and improve access. The line of the lower thoracic and upper lumbar spine is marked, as are the positions of the twelfth ribs. The angle which the ribs make with the spine will vary according to the build of the patient, but it should be borne in mind during the injection that the needle should remain inferior to the pleura. If a general anaesthetic is not employed, a local anaesthetic wheal is raised below the twelfth rib, about 7–8 cm lateral to the spine of L1. The actual distance will vary depending on the build of the patient, but the point should not be too medial, or the needle will not be able to pass antero-medially to the body of the vertebra. A 15 cm 20 or 22 gauge needle is inserted in a medial, anterior, and slightly cephalad direction, using the image intensifier to direct the needle towards the body of first lumbar vertebra (Fig. 6.3). If the needle encounters the transverse process it may be necessary to redirect the needle. In most patients, the body of the vertebra is reached at a depth of about 10–12 cm and the position on the anterolateral border of the vertebral body should be confirmed radiologically. The needle is then withdrawn slightly and redirected so that it passes anteriorly to the upper margin of L1 and comes to lie about 1 cm anteriorly to L1 and the lower margin of T12. When the needle is being inserted on the left-hand side, the pulsations of the aorta are usually sensed along the needle by the operator, when the needle approaches the correct depth. At this stage contrast medium is injected to check the plane of spread. If contrast tracks laterally, then the position of the needle should be adjusted. However, the needle may encounter one of the great vessels, resulting in the aspiration of blood. If this is inferior vena cava, then the needle should be partially withdrawn and directed more medially. If aorta is entered, then the needle can be redirected or, alternatively, advanced so as to transfix the aorta, and the injection made

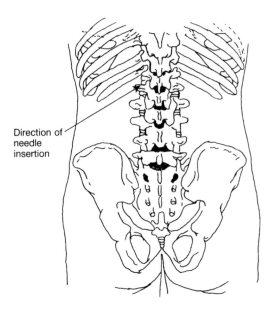

Direction of needle insertion

Fig. 6.3 The needle is inserted approximately one hand's breadth from the midline, and aimed at the anterolateral margin of T12/L1, for coeliac plexus block.

anterior to the aorta (this is the preferred technique by some (Ischia *et al.* 1983). When a satisfactory position is obtained radiologically as judged by the spread of contrast medium (Figs 6.4 and 6.5), a second needle is inserted on the other side in the same manner. Although the injection is often made with a single needle technique, results are possibly better using a bilateral approach. When both needles are satisfactorily positioned, an injection of 0.5 per cent bupivacaine 5 ml is made on either side if the patient is awake. It is then necessary to wait for about 15 minutes for the local anaesthetic to take effect, and to diffuse sufficiently to avoid diluting the neurolytic solution. If a general anaesthetic is used it is possible to proceed with the neurolytic injection. A mixture of 10–15 ml of absolute alcohol diluted with 5 ml of 0.5 per cent bupivacaine (to provide post-operative analgesia) is injected on each side after careful aspiration. Intraperitoneal or subdural injection should be avoided by the use of radiological control. The patient should then be left prone for about 20 minutes so that the alcohol can fix and reduce the risk of posterior movement of the neurolytic solution after injection. Failure to do this may increase the chances of developing a painful post-injection neuritis.

Analgesia may develop rapidly following the procedure, but in some cases takes a day or two to reach a maximum. Although the results can be very impressive, in some patients the analgesic effect wanes after a matter of

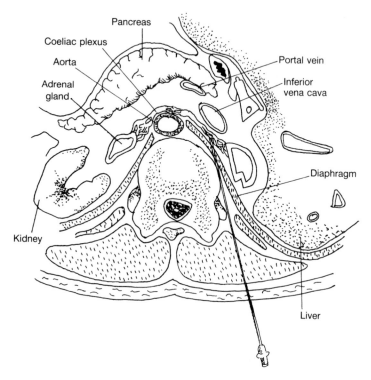

Fig. 6.4 Transverse section to show the anatomical relationships of the needle position for coeliac plexus block.

weeks or even days. There may be some case for repeating the procedure in such instances.

The main complication of this procedure, due to the extensive sympathetic blockade, is hypotension. For this reason it is essential to set up an intravenous infusion before starting the injection, and the patient's circulating volume should be maintained in this way for the following 12 hours. When the patient starts to get out of bed, usually the following morning, postural hypotension should be watched for and attaining the upright posture should be done in stages. The patient should be warned that postural hypotension may occur following the injection, and that care should be taken when getting up in the morning. The effect is more likely to be pronounced in the elderly, but it usually becomes less troublesome with time, as the circulation adapts. If the problem persists, then the use of elastic stockings may be required or, in more extreme situations, an abdominal binder. However, this situation seems to be uncommon, and the majority of

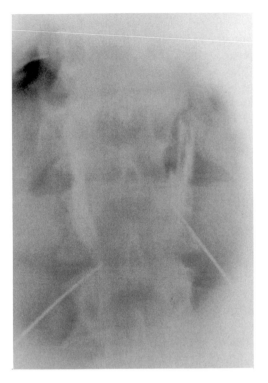

Fig. 6.5 Radiograph to show needle positions and spread of radiocontrast medium during coeliac plexus block.

patients experience little more than a transient light-headedness, and a little shoulder tip pain.

Lumbar sympathetic block

Blockade of the lumbar sympathetic chain is a commonly performed procedure for the relief of pain. Perhaps the most frequent indication is for the relief of rest pain and improvement of skin perfusion in peripheral vascular disease, but the sympathetic system has a role in maintaining peripheral pain and allodynia following nerve damage as seen in algodystrophies, and it is also responsible for much of the afferent pain sensation from the lower abdominal and pelvic viscera. Blockade of the lumber sympathetic chains is therefore indicated when pain arises from the lower abdominal or pelvic organs as a result of malignant disease, when pain is difficult to control pharmacologically. If pelvic pain is thought to be due to invasion of skeletal tissue or pelvic nerves, then a sympathetic block is unlikely to help.

The lumbar sympathetic ganglia lie along the anterolateral margins of the bodies of the lumbar vertebrae, separated from the somatic nerves by the psoas muscle and fascia. There are usually three or four ganglia connected together and lying between L1 and L5. Lateral to the chains are the psoas muscles, and anteriorly are the aorta and inferior vena cava. The retroperitoneal fascia also lies anteriorly to the chains, so that there is a potential fascial compartment into which a local anaesthetic or neurolytic solution can be injected, and be expected to spread along the chains through several segments.

The procedure can usually be easily performed under local anaesthesia, with perhaps a light sedation. It is important to have intravenous access, and a cannula should be inserted before commencing. Although for many years this block has been successfully performed without radiological control, nowadays, the use of an image intensifier is advisable, and mandatory if a neurolytic block is being performed. Whilst one can be reasonably certain of the position of the needles without X-rays, the exact plane through which the injected solution spreads can be unexpected, and spread of a neurolytic solution into the subarachnoid space or to the lumbar plexus can have disastrous results.

The patient is placed in a prone or a lateral position on the radiolucent table, and the position of the lumbar vertebral spines is determined. A wheal is raised between 5–8 cm lateral to the transverse process of L3. The actual distance will depend on the build of the patient. A 12–15 cm 20 or 22 gauge needle is then advanced, infiltrating with local anaesthetic, in a medial and anterior direction, until bone is contacted. If this occurs at about 5 cm depth, this is probably transverse process, and this can be checked radiologically. The needle is then withdrawn a little and redirected towards the anterolateral margin of the vertebral body. During this operation it is useful to have occasional anteroposterior views on the image intensifier.

When bone is contacted the needle is 'walked off' the anterolateral margin of the vertebral body, and advanced about 1 cm further.

At this stage it is necessary to check the position of the needle by the injection of contrast medium. The correct site is shown in Fig. 6.6. Although the needle may appear to be correctly positioned on the image intensifier screen, it is only possible to be sure that the correct tissue plane has been entered by the use of radiocontrast. A small volume (1–2 ml) of contrast medium is injected initially. If the needle is correctly positioned, the medium will be seen to spread in a broad 'smudge' parallel to the line of the spine, overlapping the lateral margins of the vertebral bodies, as seen in the anterposterior view (Figs 6.7 and 6.8). If the tip of the needle lies within the psoas sheath, the medium will be seen to track laterally within the psoas (Fig. 6.9),

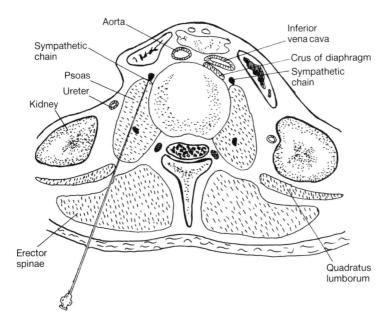

Fig. 6.6 Transverse section to show site of injection for lumbar sympathetic block.

and along the lumbar nerves; injection of neurolytic solution in this position will cause damage to the somatic nerves with consequent motor and sensory deficit. The needle should therefore be repositioned more medially, and a further injection of contrast medium checked.

If the aorta or inferior vena cava are punctured, the needle should be partially withdrawn, and repositioned. Generally no harm results, although the efficacy of the block may be reduced due to dilution of neurolytic solution in the haematoma.

If it has been decided to do a trial block, then about 10 ml of 0.25 per cent bupivacaine are injected. If, as is often the case in late malignant disease, it has been decided that the indications for a sympathetic block are sufficiently positive and a neurolytic block is to be performed, then 5–6 ml of 6 per cent aqueous phenol are injected, and a small amount of local anaesthetic or saline are flushed through the needle before withdrawal to avoid leaving a track of neurolytic solution. Some authorities recommend a larger volume of phenol, but if the needle is accurately positioned, then the smaller volumes are adequate, and help to avoid unwanted spread of the solution onto somatic nerve roots.

The patient should remain prone after the injection for about 20 minutes to avoid posterior spread of phenol onto the somatic nerves, and the blood pressure is monitored. Any hypotension is dealt with in the usual manner, with intravenous fluid replacement and, if necessary, vasopressors. If a bilateral neurolytic block has been performed, it is probably better for the patient to stay in bed overnight, and they should be warned of postural hypotension, although this is usually self-limiting.

The commonest complication apart from hypotension is a neuritis of the genitofemoral distribution, resulting in a burning sensation in the anterior thigh. This again is self limiting, but may be ameliorated by the prescription of amitriptyline.

Paravertebral block

This approach may be used to block the segmental nerve roots, primarily in the thoracic and lumbar regions. It may be useful in the control of pain arising from the body wall, especially when the pain has a peripheral origin in tissue

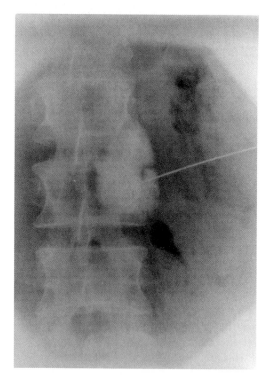

Fig. 6.7 A/P radiograph to show needle position and spread of radiocontrast medium during lumbar sympathetic block.

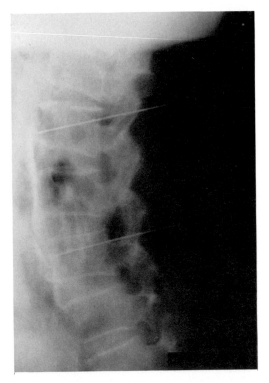

Fig. 6.8 Lateral view to show spread of contrast medium during lumbar sympathetic block.

damage, and is not primarily neurogenic. A diagnostic (prognostic) block may be performed with local anaesthetic or, with the addition of steroid, this can often be prolonged for a useful period. If a destructive block is considered appropriate, then 6 per cent aqueous phenol is the agent most commonly employed.

In the thoracic region, the tip of the spinous process is generally at a level with the intervertebral foramen of the next-lowest segmental nerve. The vertebrae are counted downwards from C7 at the base of the neck, or upwards from the level of L4 (between the top of the iliac crests). The patient is placed in the lateral position, with the side to be blocked uppermost, or occasionally it may be easier to have the patient sitting up. A 22 gauge 8 cm needle is inserted through a wheal of local anaesthetic 2–3 cm lateral to the middle of the spinous process of the vertebra below the nerve which is to be blocked. It is advanced in an anterior and very slightly medial direction until the transverse process is encountered. The needle is then slightly withdrawn,

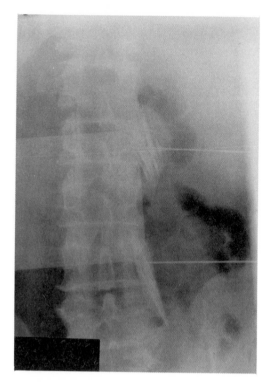

Fig. 6.9 A/P radiograph showing apparently correct positioning of needles for lumbar sympathetic block, but spread of contrast medium demonstrates that injection is spreading laterally in the psoas compartment; position is therefore incorrect.

and advanced in a slightly more cephalad direction, so that it is 'walked off' the superior margin of the transverse process (Fig. 6.10). The needle is then advanced a further 0.5 cm and, if aspiration is negative, the injection is made. There should be little resistance to injection if the needle is in the paravertebral space. If paraesthesiae are reported, the needle is withdrawn a few millimetres before injecting. A volume of 5 ml of solution will be adequate to block the segmental nerve. If (as is usual) it is desired to block several segments then, after injecting 5 ml, the needle can be withdrawn to the level of the transverse process and redirected inferiorly to that landmark before injecting to block the nerve at the next lower level.

The hazards of this procedure are mainly those of producing a pneumothorax, or of injecting a cuff of dura to result in a subarachnoid injection. The latter poses obvious risks, especially if a neurolytic block is planned. Paravertebral blocks have the disadvantage of poor prediction as to the

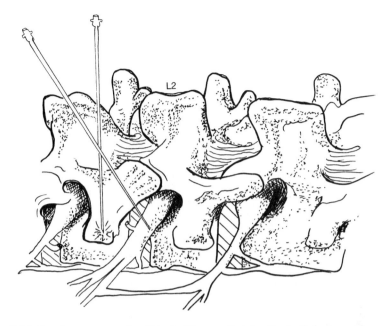

Fig. 6.10 Lumbar paravertebral block. The needle is 'walked off' the transverse process and advanced a few millimetres until a loss of resistance is encountered.

spread of the blocking solution, and many prefer to perform them under the control of an image intensifier. However, if adequate precautions are taken, the risks should be small, and paravertebral blocks can provide a simple and rapid form of pain relief on the ward for the sick patient.

Block of the lumbar nerves is similar to the procedure in the thoracic region, except that the upper border of the spinous process is on the same level as the transverse process of the same vertebra and, because of the increased depth of the intervertebral foramen, the point of skin puncture will need to be more lateral than in the thoracic region. A distance of about 4 cm from the midline is usually necessary

In patients with well-defined somatic pain of segmental distribution, if a diagnostic block in the paravertebral region has been effective with local anaesthetic, a radio-frequency thermocoagulation lesion of the dorsal root ganglion should be considered if facilities are available. It may not be possible to accurately perform such a lesion in the severely debilitated patient, but if the patient can co-operate sufficiently for accurate location of the lesion, then the procedure is safer and more effective than a neurolytic injection.

Trans-sacral nerve block

A paravertebral block of the sacral nerves can be performed by inserting a needle through the posterior sacral foramina. When considering a trans-sacral block, the anatomy of the sacrum should be studied, and special consideration given to the fact that the sacrum is the lower part of a lumbo-sacral lordotic curve, and the sacrum is also convex posteriorly.

There are four sacral foramina, anteriorly and posteriorly, through which exit the sacral nerves. The foramina lie in two straight lines which follow the triangular shape of the sacrum and approach each other inferiorly.

The patient lies prone during a trans-sacral block, and a pillow is placed beneath the hips. The posterior superior iliac spine is palpated, and a point 1 cm both inferiorly and medially to it is marked. This point should overlie the second sacral foramen. The sacral cornua are identified, and the fifth sacral foramina should lie 1 cm inferiorly and laterally to these landmarks. A line is drawn between the second and fifth foramina, and this is divided into thirds. The points at the junctions of the thirds overlie the third and fourth foramina. The first foramen lies at a point on this connecting line, about 2 cm above the second.

Needles are inserted at the points overlying the foramina, and advanced in a slightly medial direction until bone is encountered. The needle is then withdrawn, and reinserted in a progressively medial direction until the foramen is reached. For S1, the needle is advanced 1.5 cm into the foramen, for S2, 1 cm, and for S3–5, between 0.5–0.75 cm. At each site, 3–5 ml of solution are injected.

It should be remembered that blockade of the sacral nerves produces a variable spread of blockade, and blockade of S2, especially if bilateral, will result in incontinence. This is important if a neurolytic solution is injected, and a radio-frequency coagulation lesion may produce a more accurately controlled block.

Dorsal root ganglion lesion

Blockade of somatic afferent fibres can be produced with a selective destruction of nociceptive afferents by means of a radio-frequency thermo-coagulation lesion in the dorsal root ganglia (Nash 1986). This procedure has the advantage over neurolysis in that it is more selective and accurate, and has less risk of causing either a motor lesion, or a post-injection neuritis. It may also have an advantage over a more peripheral lesion by blocking afferents which can enter via the anterior roots, before synapsing in the dorsal horn. Unfortunately, the severely debilitated patient may not be able to co-operate adequately for the procedure, and as it is necessary to have any patient's help in locating the correct site of lesioning, sick patients may not be able to tolerate the discomfort involved. Nevertheless, for the control of pain which is well-localized in the distribution of a determined nerve root, and in a patient who is otherwise mobile and who has a life expectancy of a few weeks or more, then it can be a valuable procedure.

The patient is usually given a mild sedative premedication, so that full co-operation is maintained. The prone position on the operating table is usually adopted, although in certain circumstances, a lateral position may be used. Following skin preparations, a 10 cm insulated needle electrode (Radionics Sluyter-Mehta Kit 22 g cannula) is inserted through a skin wheal about 4–5 cm lateral to the midline on a level with the intervertebral foramen identified as that through which the nerve to be blocked emerges (Fig. 6.11). The needle is advanced in antero-medial direction under image intensifier control until bone is encountered (Fig. 6.12). If this is the transverse process, the needle is withdrawn slightly and redirected to pass superiorly to the process in the direction of the foramen. If the needle strikes pedicle or facetal process, then it can be 'walked off' until it slips into the foramen. Paraes-thesiae are often reported at this stage and are usually painful. The stimulator connected to the needle is then switched on at a frequency of 75–100 Hz and a potential of about 0.4 V. The needle is advanced to the posterior portion of the foramen, having checked its position both in the antero-posterior and lateral planes. The patient is asked to report the onset of tingling, and the position of the needle adjusted so that this is perceived at a potential of between 0.2–0.4 V. The patient should be asked to describe the distribution of paraesthesiae, and whether this is the same as the segmental distribution of their pain. The tip of the needle should not be advanced further than the

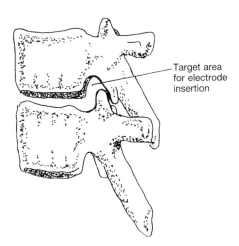

Target area for electrode insertion

Fig. 6.11 Target area for dorsal root ganglion lesion.

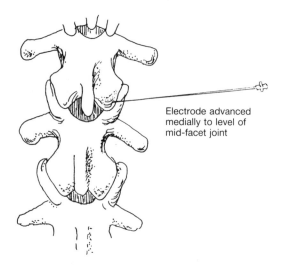

Electrode advanced
medially to level of
mid-facet joint

Fig. 6.12 Direction of needle insertion for dorsal root ganglion lesion.

midline of the facet joint, as seen on the antero-posterior (A-P) view, or the risk of dural puncture or cord damage increases considerably. The stimulation is then changed to a frequency of 5 Hz to test for motor response. There should be no motor activity at a voltage of less than twice the sensory threshold. If a motor response is obtained at low thresholds, then the position of the needle should be adjusted to a more posterior position in the foramen.

When the operator is satisfied that the needle is well-positioned close to the dorsal root ganglion of the correct nerve root, 2 ml of 2 per cent lignocaine is injected and, after a suitable interval, the lesion is made. The radiofrequency current through the needle is increased until a steady temperature of 70 °C is obtained, and this is maintained for one minute. If the needle is correctly placed, then there is no advantage in prolonging the heating phase or in repeating it. However, in order to gain adequate relief, it may be necessary to produce lesions at 2 or 3 adjacent levels. Before proceeding with radio-frequency lesioning, it is advisable to become familiar with the particular apparatus being used, and the details of the techniques as described in the handbooks for that apparatus thoroughly studied.

Subarachnoid block

Pain relief in malignant disease really entered a modern era with classic descriptions of the injection of neurolytic agents into the subarachnoid space. Thirty years ago, this was one of the most effective measures available and it rapidly gained popularity. However, as mentioned before, less destructive methods have become available and its popularity has declined, although there is still a place for the procedure in the late stages of malignant disease when pain is inadequately controlled by other means, and the potential side-effects of the method are seen as less important. The problem of sub-arachnoid block in the lumbar and sacral regions are obvious, where loss of motor function and sphincter control can be so devastating. However, in the thoracic region, the techniques can be used with fewer problems, and even cervical neurolysis has a place.

The indications for this procedure are the presence of uncontrollable pain, especially when over-sedation is unacceptable, and the pain can be located to a defined distribution of nerve roots. Especially for the lumbar and sacral areas, the technique should only be used after very careful consideration if mobility and continence are not already compromised.

The use of an image intensifier makes the procedure both more accurate and a little safer, but in desperate situations, the procedure is sometimes performed as a bedside manœuvre in the terminal stages of illness, and this may be well-justified.

Various neurolytic solutions have been advocated over the years, but the most commonly used are hyperbaric phenol 5 per cent in glycerine, and dehydrated alcohol. The latter, being hypobaric, necessitates the patient being positioned with the painful side uppermost, and this may be an advantage. However, phenol remains more popular. Its hyperbaricity tends to make prediction and control of the level of block a little more reliable, and the resulting neural blockade may be more effective. Much has been written concerning the control of the spread of neurolytic solutions in the cerebro-spinal fluid (CSF) (Mehta 1981) but it has to be stated that the various techniques used are at best only partially effective, and the risk of non-selective blockade should be considered

For subarachnoid block in the thoracic and lumbar areas, if using phenol, the patient is placed in a lateral position with the painful side dependent. Using a 20 gauge needle (if finer needles are used there is difficulty injecting the viscous phenol solution) a lumbar puncture is performed at a level corresponding to the nerve roots to be blocked. Although the vertebral level does not correspond with the cord level throughout the spine, an injection at the vertebral segmental level will produce the greatest concentration of neuro-lytic agent at the point where the posterior roots enter the dura, and this probably results in a more effective block.

It is usual to place the patient in a semilateral/supine position with the painful side lowermost. It is said that by using this position, the hyperbaric phenol will lie predominantly over the posterior roots. However, it seems unlikely that such accuracy of predicting the flow of solution is possible, and there are many supporters of the view that by positioning the patient in a straight lateral position, the patient is more comfortable, and the chances of motor blockade are no worse.

As soon as CSF is encountered, the needle is withdrawn as far as possible without losing a slight flow of CSF, to ensure that the subsequent injection is made as far posteriorly in the subarachnoid space as possible. A solution of 5 per cent phenol in glycerine is then injected, using a volume of up to 1 ml in the lumbar spine, and 1.5 ml in the thoracic region. A 1 ml syringe should be used. The needle must then be flushed with the minimal volume of saline necessary to clear it of phenol before withdrawal. It is usually suggested that the patient is then maintained in the lateral position with a slight posterior tilt for 20 minutes to allow the phenol to fix selectively over the posterior roots. However, this precaution is of dubious value, especially in the lumbar region where it seems unlikely that the phenol will be so localized in the CSF as to only bathe the posterior roots. It is probably helpful to maintain the lateral position for a period to attempt to obtain a unilateral block.

A radiocontrast medium can be injected prior to the phenol in order to predict the direction of spread of solutions in the subarachnoid space, and the table can be tilted accordingly to alter the direction of spread of a hyperbaric solution. Some authorities recommend mixing contrast with the phenol so that its spread can be viewed directly radiologically, although how much control this really provides is dubious.

For pain in the sacral distribution, much smaller volumes of neurolytic solution should be used, to minimize possible damage to sphincter control. The injection is made at L5/S1 level if possible, or if necessary at one level higher, either with the patient sitting up, or in a lateral position with the head of the table tilted upwards. Usually only 0.5 ml of neurolytic solution is employed, and if blockade is inadequate, it is usually best to repeat the procedure 2 or 3 times on subsequent days, rather than increasing the volume injected. In desperate situations where life expectancy is extremely limited, it may be justified to attempt a more widespread blockade at the first occasion.

Pituitary ablation

It was recognized long ago that some types of tumour are hormone-dependent, and destruction or removal of the appropriate source of endocrine secretion may lead to regression of the tumour. A further step was to remove the pituitary gland or destroy it with radioactive implants. This technique was widely used to slow the growth of breast and prostate metastases. In the 1960s, Moricca (1977) developed the technique of transphenoidal injection of alcohol to destroy the pituitary gland and although he called this 'pituitary adenolysis', the technique probably does not usually totally destroy the gland, but is effective in causing regression of certain tumours as well as providing pain relief. It was originally believed that the analgesia resulted from the regression of the tumour, but Moricca reported pain relief with the procedure when pain was due to widespread dissemination of other tumours which were not expected to be hormone-dependent, such as some bronchial and renal metastases. Moreover, analgesia would frequently result immediately after the procedure, or within the following couple of days; certainly before there was any clinical evidence of tumour regression, and even when the latter never occurred.

The procedure has been advocated as suitable for the relief of intractable pain resulting from any widely disseminated malignant disease. It can be performed in relatively sick patients, and can be repeated if life expectancy exceeds the analgesia produced. It is useful when pain is bilateral (unlike cordotomy), and has been described as inexpensive, although the use of an image intensifier is mandatory, and that requires a sizeable capital outlay.

However, as with any procedure, pituitary injection has its enthusiastic advocates who quote huge numbers of successful cases, but there are also potential hazards which in the light of current improved analgesic management of pain, have led to decline in the popularity of this technique in many practices.

Pituitary injection classically uses alcohol as a neurolytic agent. More recently, techniques have been described using cryotherapy (Duthie *et al.* 1983) or production of a radio-frequency thermal lesion (Zervas 1969) in the pituitary. These techniques, although requiring more specialized equipment, are believed to carry less risk than the injection of alcohol.

We prefer to perform this procedure with the patient under a light general anaesthetic, although many use sedation and local anaesthesia. The advantage of the latter is quoted as rendering the procedure suitable in the frailest of patients, but modern anaesthetic techniques may well be safer than sedation, and the procedure must be alarming for even the most stoical of patients. It is important, however, to avoid the use of opioid analgesics during an anaesthetic, as it is essential to monitor the pupillary reactions during the injection. It is also essential to protect the airway and control ventilation if the patient is anaesthetized, and a throat pack should be inserted around the endotracheal tube.

Patient selection is usually confined to those patients with widespread metastatic disease, whose pain is inadequately controlled by other means. Tumours originating from breast and prostate are especially likely to respond but, as mentioned before, other widespread metastases often respond well. If the tumour is known to be hormone-dependent, the procedure is often repeated several times in order to produce disease regression. However, in non-hormone-dependent tumours, if pain relief occurs, the procedure is not repeated until pain recurs. Patients with any infective lesion in the nose or sinuses must be excluded.

Purpose-designed pituitary needles are available. This should be a trochar and cannula about 12.5 cm long, and of 16 gauge. The internal volume is usually 0.3 ml, but this should be checked as the internal volume is important in relation to an injection of perhaps 1 ml in total; when the cannula is replaced before removal of the trochar, this internal volume is injected, and must be included in calculation of the total volume injected. It should also be given in increments of 0.1 ml as with the main part of the injection.

The anaesthetized patient is placed on the operating table with the head on a narrow radiolucent head support. This enables the operator to obtain antero-posterior and lateral views with the C arm of the image intensifier swinging around the head of the table. The head should be stabilized to prevent movement with a head ring, and the antero-posterior view checked to ensure that the head is positioned correctly in that plane. The pituitary fossa should be readily seen in the lateral view.

There is probably little point in attempting to sterilize the nasal cavity, as this appears to have little effect on infection rates. A vasoconstrictor may be useful, and cocaine paste is often applied to the nasal cavity, although this is not really necessary in the anaesthetized patient.

The trochar is inserted through the nares in a direction guided by the lateral, radiographic view, so as to aim for the posterior clinoid process. The tip of the trochar will generally pass through the wall of the sphenoid sinus with gentle pressure. It is essential to study repeated antero-posterior views to ensure that the trochar is constantly maintained in the midline. Any deviation from this plane can have disastrous consequences, and should be immediately corrected. As the needle is advanced towards the pituitary fossa the hard bone of this structure is encountered. It is usually necssary to use a small ENT hammer to tap the trochar and cannula through ths bone and into the fossa. The position (Fig. 6.13) should be repeatedly verified with both radiological views. Once through the floor of the fossa, the cannula can be positioned in different sites within the fossa for the injection. If tumour regression is the object, then a thorough destruction will involve injection at many points within the gland on several different occasions. For analgesia, deposition of alcohol in two separate sites within the fossa may suffice. The volume of absolute alcohol rarely exceeds 1 ml given from a tuberculin syringe, and that should include the internal volume of the cannula as mentioned above. The injection is given in increments of 0.1 ml. After each increment, a period of 2 minutes is allowed, and the pupil size and reaction to light is noted. If there is any change in size, or the reaction to light becomes sluggish, then if the change does not revert after 5 minutes, the procedure is abandoned until another day. This precaution should prevent one of the major hazards of the procedure: the optic chiasma bears a close relation to the pituitary fossa and spread of alcohol beyond the fossa may damage visual acuity or interrupt the light reflex arc. Damage to the occulomotor nerves is

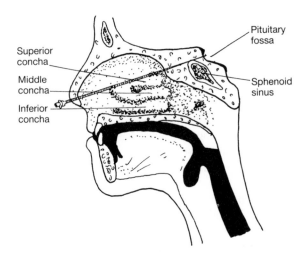

Fig. 6.13 Needle position for injection of pituitary fossa. During the advancement of the needle, its position is repeatedly checked in the A/P view to confirm that it remains in the midline.

also a possible complication. Damage to vision or diplopia would be an unacceptable complication and although if precautions are taken to check pupil size and reaction, the complication is rare, its occasional occurrence is one of the major drawbacks of this procedure. It is an unresolved problem as to whether obtaining informed consent extends to informing the patient that they may lose their sight, and thus perhaps denying the patient a worthwhile method of relieving severe pain because of an uncommon event.

Although this technique has a relatively low mortality, considering the likely condition of the patients being treated, there are a number of important complications which may arise. Headache is perhaps the least serious of these. It commonly occurs in the post-injection period, but is probably more common when the procedure is performed under local anaesthesia. It is generally self-limiting.

Rhinorrhoea may develop following pituitary injection. Although this usually only lasts for a day or two, it can persist for longer periods, but rarely needs any corrective treatment as healing will eventually occur in the vast majority of cases. Some would recommend the prescription of a sulphonamide antibiotic during the period of rhinorrhoea.

The most likely complication seen after pituitary injection is diabetes insipidus. This seems to ocur after most injections, but in the majority of cases it diminishes gradually over a few days. Providing that adequate hydration can be maintained, no further action is required in most patients. If the condition persists for longer, or it becomes difficult to keep up with fluid losses, it may be necessary to administer antidiuretic drugs. Desmopressin nasal drops (DDAVP) can be given 1–4 times a day until the condition begins to resolve, and only occasionally is long-term maintenance required. Piton snuff is an alternative preparation, and severe cases may necessitate the use of vasopressin tannate injections.

Other pituitary hormones may be affected, and it is usual to cover patients for a few weeks with oral steroid supplements (e.g., prednisolone 2 mg twice daily) and then pituitary function reassessed. Mineralocorticoids may be required and in a few cases thyroxine may be necessary. The latter is not required initially, and it is advisable to check the level of thyroid function after a month.

Neurological complications are fortunately uncommon, but can have major consequences. Hyperpyrexia, hyperphagia, and lethargy have all been reported, and presumably are due to damage to the hypothalamus. Occular defects are of more concern. As mentioned above, the site of injection is close to the optic chiasma, and visual field defects can result. This risk is minimized by taking appropriate precautions, monitoring pupil size and light reflex during carefully graduated stages of the injection. With time, visual field defects will often improve, but a permanent defect can result, and complete blindness has occurred in some cases. Damage to occulomotor nerves can cause diplopia or strabismus.

The way in which pain is relieved by injecting the pituitary is unknown. As discussed previously, the original belief that it was entirely due to tumour regression does not fit with observations. Another theory suggested that analgesia resulted from an effect of the injected alcohol on surrounding structures, particularly the hypothalamus. Injection of contrast medium with the alcohol certainly produced radiological evidence of spread to several areas, including the third ventricle. However, cryolesioning and radio-frequency thermocoagulation in the pituitary fossa, where extension of the lesion is not likely, are increasingly preferred as they produce equally as good results as alcohol injection but with much less risk of causing neurological damage. It is possible that endogenous opioids are involved in the production of analgesia and it has been suggested that pituitary destruction prevents the inhibition of their production, but the evidence is conflicting and the mechanism is still one of speculation. Recent work on the production of pain relief by electrical stimulation of the pituitary may lend support to the idea of humoral factors being involved.

In summary, partial damage to the pituitary gland, traditionally by the injection of alcohol, or more recently by cryolesions or thermal lesions, will produce good analgesia in about a third of patients with widespread intract-able pain in malignant disease, and partial relief in another third. However, the morbidity, particularly with alcohol injection, and the improvement of other methods of analgesia have led to some decline in use of the method. It will probably continue to have a place, particularly for hormone-dependent tumours, and the use of cryolesions or thermal lesions will improve the safety of the technique.

Anterolateral cordotomy

One of the most effective methods of providing pain relief in malignant disease is by the interruption of the anterolateral tracts in the spinal cord. This was originally developed as a surgical procedure involving cutting of the tract under direct vision. Percutaneous techniques are now widely used which, because of the reduced morbidity and mortality associated with this approach, have made the procedure more widely applicable and available for certain types of pain. As with many of the more complex procedures, the technique should ideally be learnt from an experienced operator, and written descriptions cannot be adequate to enable the novice to practise the procedure without some previous practical experience. However, it is important to have an understanding of the procedure, so that cases can be appropriately referred if the necessary personal experience is not available. The practical details of this procedure will therefore only be outlined in this chapter. For more detail, the reader is referred to the excellent account of Lipton (1979).

From the anatomy of the spinal tract (Fig. 6.14), it would be expected that section of the anterolateral spinothalamic tracts should produce analgesia, plus loss of temperature sensation on the contralateral side of the body below the lesion. A proportion of pain fibres do not appear to cross over in the spinal cord, but the majority will be involved in a lesion in the anterolateral quadrant of the cord. In reality, a larger area than the spinothalamic tract, involving other tracts, is probably involved in the lesion.

There are two major limitations to anterolateral cordotomy as a method of providing analgesia. Firstly, because a lesion is not generally performed bilaterally due to the unacceptable complications, the pain has to be unilateral, and somatic. Secondly, the procedure is unsuitable for patients with a life expectancy of more than about a year because the 'plasticity' of the central nervous system (it is not like a permanently 'hard-wired' system) means that changes occur within the spinal cord following a codotomy, just as they do after any neurodestructive event, and pain eventually returns. It may be a different type of pain from that which was originally experienced, but it is usually unpleasant and refractory to treatment. For this reason, there are few who advocate the use of cordotomy for patients with benign pain and a normal life expectancy although, given these limitations, it is undoubtedly one of the most effective methods of relieving pain.

Centres with a large experience of cordotomies will inevitably be referred

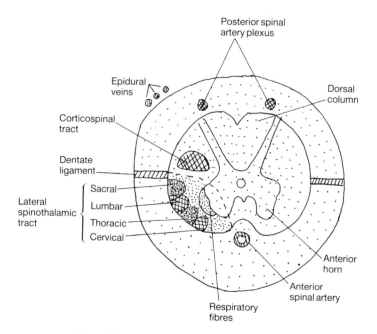

Fig. 6.14 Location of the spinothalmic tract.

large numbers of patients for the procedure but, with the increasing effective-ness of other forms of analgesia, the number of patients who fulfil the necessary criteria and in whom pain cannot be adequately controlled by other means is relatively small in the average district hospital. It is therefore questionable whether the expense of the necessary equipment and the relat-ively low level of expertise that can be maintained by the occasional practitioner justifies such procedures outside regional centres.

A reasonable degree of respiratory function is necessary in patients selected for cordotomy, as occasionally the procedure can produce a hemi-diaphragm. Also, the patient needs to be capable of maintaining co-operation under light sedation. It is usual to premedicate the patient, and the sedative drug chosen should be one of those with which the operator is familiar. It is important not to oversedate, as the patient's co-operation is essential for both the safety and the success of cordotomy.

The position of the patient is generally supine with the head flexed and held in a head holder, usually purpose made, and designed to maintain the head in a constant and comfortable position during the procedure. A C-arm image intensifier is placed at the head of the table so that it is free to rotate around the patient's head and neck. Skin infiltration with local anaesthetic is performed 1 cm below and posterior to the mastoid process. Using infiltration with local anaesthetic, an 18 gauge spinal needle is advanced towards the posterior border of the cord, between C1 and C2. Blood may drip from the needle as it penetrates veins in the epidural space. The needle is then advanced a little further until the dura is penetrated and CSF is obtained, and the tip of the needle can be seen to lie midway between the medial border of the pedicles on the lateral side and the lateral border of the odontoid process on the medial side. At this stage the advance is stopped, and a little CSF is withdrawn and mixed with radiocontrast medium to produce an emulsion. When 1–2 ml of this emulsion are injected, it should be possible to see three separate lines on the image intensifier; the posterior border of the dura, the dentate ligament, and the anterior border of the cord. At this stage a thin electrode can be introduced through the needle, so that its tip protrudes 2 mm from the end of the needle. The electrode is advanced towards the cord, aiming 1–2 mm anterior to the dentate ligament for the lower fibres, and slightly more anterior for the upper pain fibres. The needle–electrode combination is advanced until there is a sharp rise in electrical impedance, which indicates the correct depth of penetration.

Once the electrode has penetrated the cord, more accurate positioning must be achieved with the use of stimulation testing. Motor stimulation is tested first to ensure that the electrode is not in the corticospinal tract. Although some small movements of the face and trapezius muscles may occur, it is important to check that there is no movement of the limbs on the ipsilateral side when stimulation is performed at a frequency of 2 Hz and a potential of around 1.5 V. Any muscle activity indicates incorrect placement, and the electrode must be withdrawn and directed more anteriorly. Sensory stimulation is then tested at a frequency of 100 Hz, but at a much lower voltage. The potential is increased from zero until sensory stimuli are perceived by the patient. The threshold for these should be less than 0.4 V, and they should normally be experienced on the contralateral side of the body. Testing must be meticulous if analgesia is to be produced in the correct area with minimal morbidity.

The coagulation lesion is usually produced in stages, with current passing for periods of 10, 20, and then 30 seconds. Each period is followed by testing of ipsilateral hand grip and leg raising. A maximum of 80 °C for 60 seconds in total is used for the coagulation.

There are many variations in technique for this procedure, depending on the operator and the equipment used. The description given here is a bare outline of the procedure, and expertise should be obtained by personal tuition.

The procedure of cordotomy is not without hazard, and many complications have been described. Unwanted damage to the cord can occur, with various neurological sequeliae. Some motor weakness is common; ataxia and dysaesthesia are less common. Erectile impotence has been reported in males. Urinary problems can occur, generally only after bilateral cordo-tomies. Respiratory insufficiency may occur after a unilateral lesion, but generally only if there was a pre-existing reduction in respiratory capacity. Neck pain and headache may occur after the lesion. A complication that was originally unexpected was the phenomenon of opioid overdose in patients who had been maintained on opioid analgesics prior to a cordotomy. Following lesion production, the removal of the physiological opioid antagonist effect of pain caused a relative toxicity of opioids, even though the dose had not been increased.

The results of cordotomy are good, and in correctly selected cases analgesia can be expected in between 80–95 per cent of patients, depending on the experience of the operator. The potential morbidity and the limited duration of analgesia mean that the procedure has only a limited application in non-malignant conditions or indeed, in the view of many, is never appropriate in such situations.

Radiotherapy

The practice of radiotherapy is obviously beyond the scope of this book, but the pain therapist may often require the services of a radiotherapist. In this context, radiotherapy may be used palliatively to relieve pain especially where bone invasion, nerve plexus invasion, lymph node masses, or invasion of the chest wall is a source of intractable pain. Palliation is often possible, even though the tumour is designated 'radio-resistant'. There are many advantages in using low-dose, single-exposure treatment for relief of pain; sick patients will not require repeated transport to the radiotherapy centre, and the systemic effects will be reduced. It also means that normal tissue will be less damaged, and healing can occur more readily after treatment. As there is no evidence that pain relief is related to the dose of radiotherapy or to evidence of tumour regression, such treatment seems to be appropriate.

For pathological fractures of long bones (or impending fractures) radiotherapy alone is unlikely to remove the pain. The problem is best dealt with by surgical fixation, followed by radiotherapy. Good fixation is more difficult to achieve if attempted after radiotherapy.

As with most treatments, radiotherapy is not without complications. Apart from the immediate systemic effects, especially on the gastrointestinal and erythropoietic systems, pain can arise secondarily from irradiation damage to normal tissues, especially mucous membranes, resulting in painful proctitis or oropharyngeal lesions. In the longer term, radiation neuropathy may present an intractable problem.

Other neuroablative procedures

Apart from cordotomy, other neuroablative procedures are sometimes indicated for the relief of pain in cancer. Indications are uncommon these days and, of course, access to a neurosurgical service is necessary. Nevertheless such procedures are occasionally useful, especially in treating pain from tumours of the head and neck, which may be less accessible to other forms of analgesia.

Implantable drug delivery systems

There has, in recent years, been a vogue for the use of implantable drug delivery systems. Since it was shown that the injection of opioid drugs into the epidural or intrathecal compartments could produce analgesia, the merits of providing medication by these routes have been heatedly discussed. An epidural catheter can be introduced and tunnelled subcutaneously for long-term injection, either intermittently or by infusion. Another development is to implant subcutaneously a device which will deliver a slow infusion or receive intermittent percutaneous injections for delivery by the epidural route. Although some patients may tolerate opioid therapy via the epidural route better than orally, as the doses used are considerably lower, there is little hard evidence to support the belief that superior analgesia is possible by this route. In any case, it would seem prudent to give oral analgesics an effective trial before contemplating such an invasive (and very expensive) procedure. In a few patients it may prove to be the only way to achieve effective opioid analgesia, especially if there is poor tolerance to side-effects when using the oral route.

A summary of the various options available for controlling pain is given by Fig. 6.15.

The final days

Often when the patient with cancer is nearing the end of life, the administration of medication presents new problems. If consciousness is impaired, the use of oral medication will not be practical. Some drugs will cease to be useful at this stage, whilst it is important to continue some drugs as long as life continues.

Analgesia should generally be continued. Cessation may produce unpleasant withdrawal problems, and even the semiconscious patient should be kept pain-free because, although their communication with those around may be failing, it is difficult to know what sensations the dying patient is experiencing.

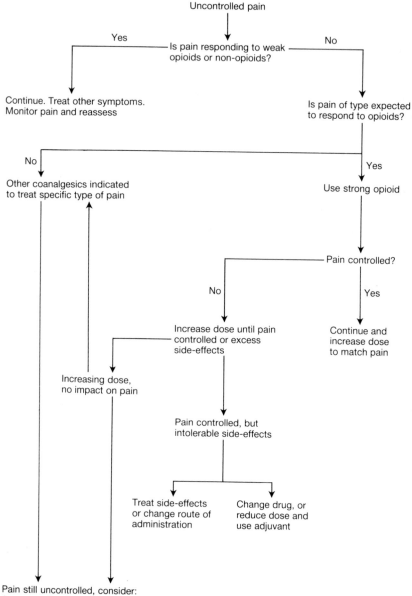

Uncontrolled pain

Yes — Is pain responding to weak — No
opioids or non-opioids?

Continue. Treat other symptoms.
Monitor pain and reassess

Is pain of type expected
to respond to opioids?

No

Other coanalgesics indicated
to treat specific type of pain

Yes

Use strong opioid

Pain controlled?

No

Yes

Increase dose until pain
controlled or excess
side-effects

Continue and
increase dose
to match pain

Increasing dose,
no impact on pain

Pain controlled, but
intolerable side-effects

Treat side-effects
or change route of
administration

Change drug, or
reduce dose and
use adjuvant

Pain still uncontrolled, consider:
1. aggravating factors, e.g., fear, insomnia
2. other physical symptoms increasing distress
3. nerve blocks
4. surgery
5. radiotherapy

Fig. 6.15 Summary of possible actions against uncontrolled pain.

For this reason it is usual to continue opioid analgesics by subcutaneous infusion, if necessary. Other weak oral analgesics can be stopped and, if necessary, an opioid substituted.

Anti-emetics, anticonvulsants, and psychotropic drugs can be continued by infusion, injection, or suppository. Other drugs acting on the cardiovascular, respiratory, and gastrointestinal systems can generally be discontinued. However, if drying secretions in the pharynx cause the dying patient to produce respiratory noise ('death rattle') which is distressing to relatives, a subcutaneous dose of 0.4 mg of hyoscine may be helpful.

Little has been said in this chapter about the psychological, social, and perhaps spiritual management of the patient in pain from malignant disease, although these are undoubtedly important aspects of the care of such patients. Even less has been said about the management of those caring for the dying patient. These points emphasize the fact that care of the dying should involve a range of persons with a variety of skills and aptitudes, so that these important aspects can be given attention as well as the relief of the physical symptoms of the dying patient.

References

Baines, M. J. (1978). Control of other symptoms. In *The management of terminal disease* (ed. C. H. Saunders), pp. 99–118. Edward Arnold, London.

Baines, M., Oliver, D. J., and Carter, R. L. (1985). Medical management of intestinal obstruction in patients with advanced malignant disease. *Lancet*, Nov. 2, 990–3.

Bonica, J. J. (1985). Treatment of cancer pain: current status and future needs. In *Advances in pain research and therapy*. Vol. 2 (ed. H. L. Fields *et al.*), pp. 589–616.

Duthie, A. M., Ingram, V., Dell, A. E., and Dennett, J. E. (1983). Pituitary cryoablation. *Anaesthesia*, **38**, 448–51.

Evans, P. J. D. (1981). Narcotic addiction in patients with chronic pain. *Anaesthesia*, **36**, 597–602.

Foley, K. M. (1979). Pain syndromes in patients with cancer. In *Advances in cancer research and therapy*, Vol. 2 (ed. G. Klein and S. Weinhouse), pp. 59–75. Raven Press, New York.

Hindley, A. C., Hill, E. B., Leyland, M. J., and Wiles, A. E. (1982). A double blind controlled trial of salmon calcitonin in pain due to malignancy. *Cancer Chemotherapy Pharmacology*, **9**, 71–4.

Ischia, S., Luzzani, A., Ischia, A., and Faggion, S. (1983). A new approach to neurolytic block of the coeliac plexus: the trans aortic technique. *Pain*, **16**, 333–43.

Lipton, S. (1979). *Relief of pain in clinical practice*, pp. 126–78. Blackwell, Oxford.

McQuay, H. J., Moore, R. A., Bullingham, R. E. S., Caroll, D., Baldwin, D., Allen, M. C. *et al.* (1983). High systemic relative bioavailability of oral morphine in both solution and sustained released formulation. In *Advances in morphine therapy*, Royal Society of Medicine International Congress and Symposium Series, No. 64

(ed. E. Wilkes and J. Levy), pp. 149–64. Royal Society of Medicine Services Ltd, London.

McQuay, H. J., Caroll, D., and Moore, R. A. (1988). Postoperative orthopoedic pain—the effect of premedication and local anaesthetic blocks. *Pain*, **33**, 291–6.

Mehta, M. (1981). Improvements in spinal injection treatment for cancer pain. In *Persistent pain: modern methods of treatment*, Vol. 3 (ed. S. Lipton and J. Miles), pp. 265–79. Academic Press, London.

Moricca, G. (1977). Pituitary adenolysis in the treatment of intractable pain from cancer. In *Persistent pain: modern methods of treatment*, Vol. 1 (ed. S. Lipton), pp. 149–73. Academic Press, London.

Morton, A. R., Cantrill, J. A., Pillai, G. V., McMahon, A., Anderson, D. C., and Howell, A. (1988). Sclerosis of lytic bone metastases after disodium aminohydroxypropylidene biphosphonate (ADP) in patients with breast carcinomas. *British Medical Journal*, **297**, 772–3.

Nash, T. P. (1986). Percutaneous radiofrequency lesioning of dorsal root ganglia for intractable pain. *Pain*, **24**, 67–73.

Oliver, D. J. (1985*a*). The use of the syringe driver in terminal care. *British Journal of Clinical Pharmacology*, **20**, 515–6.

Oliver, D. J. (1985*b*). The use of methotrimeprazine in terminal care. *British Journal of Clinical Practice*, **39** (9), 339–40.

Oliver, D. J. (1988). Syringe Drivers in palliative care: a review. *Palliative Medicine*, **2**, 21–6.

Regnard, C. F. B. and Davies, A. (1986). *A guide to symptom relief in advanced cancer*. Haigh & Hochland, Manchester.

Sawe, J., Dahlstrom, B., Paazlow, L., and Rane, A. (1981). Morphine kinetics in cancer patients. *Clinical Pharmacology Therapeutics*, **30**, 629–35.

Twycross, R. G. (1974). Clinical experiences with diamorpine in advanced malignant disease. *International Journal of Cinical Pharmacology. Therapy and Toxicology*, **9** (3), 184–98.

Twycross, R. G. (1977). Choice of strong analgesic in terminal cancer: diamorphine or morphine? *Pain*, **3**, 93–104.

Twycross, R. G. (1982). Morphine and diamorphine in the terminally ill patient. *Acta Anaesiologica Scandinavica Supplementum*, **74**, 128–34.

Twycross, R. G. and Lack, S. A. (1983). *Symptom control in advanced cancer: pain relief*. Pitman, London.

Twycross, R. G. and Lack, S. A. (1986). *Symptom control in advanced cancer: alimentary symptoms*. Pitman, London.

Walsh, T. D. (1983). A controlled study of MST Continus tablets for chronic pain in advanced cancer. In *Advances in morphine therapy*, Royal Society of Medicine International Congress and Symposium Series, No. 64 (ed. E. Wilkes and J. Levy), pp. 97–102. Royal Society of Medicine Services Ltd., London.

Walsh, T. D. (1984). Oral morphine in chronic pain. *Pain*, **18**, 1–11.

Zervas, N. (1969). Stereotactic, radiofrequency surgery of the normal and the abnormal pituitary gland. *New England Journal of Medicine*, **280**, 429–37.

7 Neurogenic pain

Classically, pain has been conceived as resulting from the stimulation of peripheral nociceptors, due to actual or potential tissue damage. The resulting stimuli are transmitted to, and interpreted by, the central nervous system (CNS) to produce an appreciation of and localization of pain. This pain can be controlled by the use of analgesic drugs, or by the blockade of afferent transmission with the use of nerve blockade or destruction. However, experience of patients with chronic pain immediately suggests that this situation frequently does not fit the symptoms and response to treatment shown in many patients suffering from chronic pain. Pain may arise within the nervous system, centrally or peripherally, as a result of damage or malfunction of a part of that system, without any receptor stimulation being involved. This is what is commonly known as neurogenic pain, and it is possible that in a third or more of patients presenting with chronic pain, neurogenic pain is, at least, a component of the pain experienced in those patients. This type of pain seems to be distinct from pain of other origins, either that from nociceptor stimulation, or even from pain resulting from the nociceptive activity within peripheral nervous tissue. These differences have important consequences for the management of such pain.

The complexity of the CNS is gradually becoming more apparent, and it is increasingly evident that the elegant anatomical and functional mapping of the classical neurophysiologists is only a superficial representation of the truth. The CNS is no longer regarded as a 'hard-wired' system, where specific receptors make permanent contact with specific areas of the brain via fixed tracts. It is true that certain regions and tracts appear to be associated with particular functions, but the connections within those areas are by no means permanent, and the function of specific neurons can adapt and change under variable stimuli. Nociceptive afferent stimuli may sensitize neighbouring cells within the dorsal horn of the spinal cord (wide dynamic range neurons) so that cells which were previously acting as mechanoceptors may now function as nociceptors, and result in irradiation of pain over a wider area than that in which the injury originally occurred. When a peripheral nerve is damaged, up to 50 per cent of afferent C fibres can develop mechanosensitivity, resulting in widespread tenderness surrounding the area of injury. These fibres may also develop chemical sensitivity, so that they have enhanced responses to noradrenalin and sympathetic stimulation.

Following damage to a peripheral nerve, there is intense chemical activity within the dorsal root ganglion, and axonal transport of the chemical mediators produced may occur in both directions. The dorsal root ganglion itself may become the source of aberrant discharges, and dorsal horn cells may become altered so that they develop an enhanced response to normal outputs and begin to respond to new inputs. In some way it seems that when peripheral nerves are damaged, there may be increased activity of the stimulated sympathetic system which can itself then stimulate nociceptor discharge, and possibly the development of more receptors for noradrenalin.

Following injury, repeated bombardment of the dorsal horn neurons with nociceptive impulses may result in enhanced activity of 'wind-up' in these cells, producing a pain memory which probably involves several receptor and transmitter systems. Prevention of this wind-up before it results in the development of chronic pain may explain why provision of analgesia prior to nerve destruction, as in limb amputation, may reduce or prevent the development of chronic pain syndromes (McQuay and Dickinson 1990; McQuay et al. 1988; Bach et al. 1983).

Neurogenic pain can result from a wide range of causes of damage to the CNS or peripheral nervous system and, although each syndrome has its own particular characteristics, they have certain features in common which distinguish the pain as having an origin in nerve damage. There is nearly always a sensory deficit, or distortion. This deficit may be apparent as a reduction in activity of large afferent fibres, so that light touch and vibration sense may be diminished. It may be that normal stimuli are felt, but the sensation is a strange and often unpleasant feeling; this is termed dysaesthesia. Normally non-painful stimuli may be perceived as painful in the affected area, allodynia, or normally painful stimuli may be excessively painful, hyperpathia. Patients experiencing neurogenic pain may complain of abnormal sensitivity in the area of skin affected, burning pains, and sudden lancinating pains, often described as 'shooting pains'. Patients frequently find it impossible to describe verbally the painful sensations that they experience from a neurogenic origin as, unlike post-traumatic nociceptor pain, it is not a common experience of the healthy.

A common accompaniment to neurogenic pain is an autonomic dysfunction. This may be represented as a full autonomic dystrophy, with dystrophic changes in skin, bone, and muscle along with the typical burning dysaesthetic pain of causalgia. In other cases, there may simply be minor indications of sympathetic overactivity, with excess sweating in the affected limb, changes of skin colour from white to purple, and minor swelling and shininess of the skin and subcutaneous tissue. Pain associated with autonomic dysfunction may typically be exacerbated by cool temperatures or emotional upsets. The affected skin is often painfully hypersensitive, so that even normal contact with light clothing becomes unbearably painful.

The main feature of neurogenic pain which influences management is that it does not generally respond to what are normally regarded as analgesic drugs, including opioids. This perhaps emphasizes the concept that neurogenic pain is not concerned with the normal nociceptive pathways and transmitters. Many patients with pain of neurogenic origin needlessly consume large quantities of potent analgesics with no relief of their pain. However, there are exceptions, and occasionally some of these patients have a component of their pain, perhaps a constant background aching sensation, which may respond partially to opioids; there is in rare situations a place for carefully assessed and monitored opioid analgesia where this can be demonstrated to give partial relief. Nevertheless, the burning dysaesthetic pains, the lancinating pains, and allodynia do not respond to this group of drugs and should be treated by other means.

Neurogenic pain can result from central or peripheral damage to the nervous system, although in some instances it is not entirely clear what is the nature of the site of the damage, and the term is applied because of the presenting nature of the pain symptoms. Central neurogenic pain may follow vascular damage in the brain or spinal cord, or occasionally the damage may be iatrogenic. An example of the latter is seen following certain pain-relieving destructive procedures. The technique of anterolateral cordotomy involves the percutaneous or open destruction of anterolateral spinothalamic tracts in the spinal cord to produce analgesia below the lesion in the contralateral side of the body. This is normally only done in patients with malignant disease who have a limited life expectancy. As mentioned before, the nervous system is a 'plastic' system, and is capable of adapting to injury, often overcoming the block produced. In time, many patients who have had their spinothalamic tracts sectioned develop an unpleasant neurogenic type of pain which may be worse than the original pain for which they were treated. Unfortunately, the new pain is unresponsive to treatment. Similarly, many patients suffering from trigeminal neuralgia are subjected to destructive lesions of the Gasserian ganglion. This commonly relieves the neuralgia, but in a few cases results in an area of complete anaesthesia, associated with an intense burning pain, which is resistant to treatment (anaesthesia dolorosa). Unfortunately, there are many examples of neurogenic pain which result from well-intentioned destructive procedures aimed at relieving pain.

Central neurogenic pain often results from a cerebrovascular event and, although the resulting damage does not always involve the thalamus, such pain is usually labelled as the 'thalamic syndrome'. This may come on immediately following a stroke, or often develops over the succeeding months as a constant burning pain and hyperpathia over one side of the body. This may affect just the face, or a limb, or may even affect the whole of one side of the body.

Peripherally, neurogenic pain can result from a variety of infectious, metabolic, or traumatic events. The pain following nerve root or plexus

avulsion, peripheral neuropathies, post-herpetic neuralgia, and post-amputation pain are some common examples, which will be discussed in further
detail. However, as the aetiology of these conditions is poorly understood,
whether the pain is truly peripheral or central in origin is a subject of conjec-
ture, and the distinction between the two is inevitably blurred.

Management of neurogenic pain

Considering the above, it can be expected that the control of neurogenic pain
involves different principles from the management of nociceptive pain.
Opioids and other analgesics are unlikely to be very effective (Arner and
Meyerson 1988) and the destruction of neural tissue to correct a problem
resulting from previous damage or malfunction of that system rarely seems
appropriate. In fact, destructive procedures may in the long term aggravate
the situation. One of the few ways in which we can influence the behaviour of
the CNS following injury is by the modulation of afferent stimuli, and when
destructive procedures are performed this option of treatment is being
destroyed. It seems that to influence neurogenic pain, it is necessary to
modulate abnormal neural activity within the CNS, either by altering the
afferent stimulation or by influencing neurotransmission systems within that
system.

The anticonvulsants

This group of drugs, introduced for the control of epileptiform conditions,
have found an important role in the control of pains which have a 'shooting'
quality, and which are generally believed to be a result of nerve damage or
malfunction (Swerdlow 1986, 1980). The nature of the pain suggested that
trigeminal neuralgia may be responsive to drugs which stabilize nerve
membrane, and carbamazepine was found to be extremely effective in
controlling this condition, such that response to the drug almost became a
diagnostic feature of trigeminal neuralgia. The exact mode of action of these
drugs is not known, but they can be effective in a wide range of painful conditions which are characterized by a sharp shooting or lancinating type of pain.
This may be by a membrane-stabilizing effect, causing a general reduction in
hyperexcitability of neurons, or it may be that anticonvulsants alter neurotransmitter activity. It is known that carbamazepine can increase brain levels
of 5-hydroxytryptamine (5-HT), and sodium valproate can increase levels of
gamma-aminobutyric acid. This latter compound may be responsible for
inhibiting abnormal discharges in nervous tissue.

The most widely used anticonvulsant for managing pain is carbamezepine.
This has established itself as the first-line drug for treating trigeminal

neuralgia. To be effective, the drug must be taken on a regular basis, and the effective dose range can vary widely between individuals. A usual starting dose is 100 mg three times daily, although in some elderly patients it may be necessary to start as low as 100 mg, at night. The dose is then gradually increased until an effective dose is achieved, or side-effects become intolerable. The effective dose may need to be as high as 400 mg three times daily, but at this dose side-effects of sedation, ataxia, and gastrointestinal upset are common. Rarely, thrombocytopenia and bone marrow depression may occur, and patients on long-term therapy should have occasional blood counts. Although carbamazepine can be a most effective drug, its use may be limited by poor tolerance and, in addition, pain may begin to reappear after a time. This tolerance may be due to increased metabolism of the drug and demands either an increase in dose, or consideration of alternative therapies. Patients who have been on long-term anticonvulsants should be withdrawn gradually if the therapy is changed, to reduce the likelihood of epileptic attacks.

Sodium valproate has become the first choice of anticonvulsant drug for treating lancinating pain in many centres, although not perhaps as widely used as carbamazepine. This drug appears to be less toxic than the latter, and is generally better tolerated. A starting dose of 100–200 mg three times daily is usual, and gastric irritation can be reduced if the drug is taken with food. As with carbamazepine, the dose is increased gradually in line with response and side-effects. Gastrointestinal upset is the commonest side-effect. Occasionally, dizziness and sedation may be troublesome although this seems to be less common than with carbamezepine. Alopecia is occasionally experienced, although this is reversible on stopping the drug. Disturbances of hepatic function may occur during treatment with valproate, and this can be detected by routine measurement of liver enzymes. Whether the detection of mildly raised liver enzymes justifies withdrawing the drug is a debatable point. Again, high doses may be required to gain control of lancinating pain, and this may necessitate increasing the dose up to 600 mg three times daily.

Phenytoin has been used to control shooting pain, and its membrane-stabilizing properties suggest that it should be particularly successful in this way. However, it has a narrow therapeutic range, and the development of skin reactions, nystagmus, and gastrointestinal problems make it less useful than valproate. It is occasionally useful as a second- or third-line drug, where the other drugs have been unsatisfactory. Doses rarely exceed 100 mg three times daily.

Clonazepam is a benzodiazepine primarily used for treating epilepsy, but again these properties render it a useful drug in the management of lancinating pain. Although it appears to be relatively non-toxic, clonazepam produces excessive sedation in many patients, and frequently it is impossible to reach a therapeutic dose. It is best to start at a dose of 0.5 mg at night and, if this is tolerated, increase the dose by 0.5 mg every few days. A response is usually apparent by the time a dose of 2 mg has been reached, if it is to be effective, but occasionally it is necessary to increase this to 4 mg if this can be tolerated. The fact that clonazepam is a benzodiazepine raises worries about the effects of long-term use. Because of this, it is rarely a first choice of drug for this type of pain, but occasionally it will be found effective where the other anticonvulsants have not been, and its use is then justified. The patient should, however, be aware that sudden cessation of the drug should be avoided.

Other anticonvulsants and membrane stabilizers have been reported as being effective in the management of lancinating pain. Clobazam and flecainide are two which are occasionally used in this way, although there is as yet little data on the use of these drugs in pain control.

When dealing with lancinating, shooting pain, it may be necessary to try the effect of more than one anticonvulsant drug sequentially, as individual tolerance and the response of the pain vary considerably between individuals. Most would consider sodium valproate or carbamazepine as the first-choice drug, and then proceed through the others if these were not satisfactory. The anticonvulsant drugs are often prescribed in conjunction with other 'centrally-acting' drugs especially with the tricyclic antidepressants, as neurogenic pain is commonly a complex of different pain modalities which requires more than a single drug management.

Tricyclic antidepressants

The tricyclic antidepressants have a valuable role in the management of chronic pain (Lee and Spenser 1977), and it is assumed that this is related to their ability to reduce the re-uptake of catecholamine neurotransmitters. These drugs may also elevate levels of 5-HT and it has been proposed that these two effects of tricyclic antidepressants may act to enhance the activity of descending inhibitory tracts within the CNS, and so modify activity within the dorsal horn of the spinal cord. Drugs such as zimelidine which, when it was available, was claimed to be a highly specific 5-HT uptake inhibitor, appear to be no better, though perhaps as good, analgesics as the 'dirty' drugs like amitriptyline, and the pharmacology of their analgesic action must be more complex than such relatively simple explanations suggest. There is some evidence to suggest a relationship between the therapeutic efficacy of these drugs and changes in the levels of endorphin and indolamine activity in chronic pain states (Johansson *et al*. 1980).

Apart from any benefits that these drugs may have on the depression which frequently accompanies chronic pain, it appears that they have an intrinsic analgesic activity which is particularly useful in managing the type of pain resulting from nerve damage. The burning dysaesthetic type of pain seen in post-herpetic neuralgia is the pain most frequently treated with tricyclic antidepressants, although they also find a place in the treatment of certain types of intrinsic pain, such as atypical facial pain, where no neuropathology can be demonstrated. It may be that in these idiopathic pain syndromes there is an underlying depressive condition responsible for the pain, but this cannot usually be demonstrated. It is apparent, therefore, that tricyclic antidepressants have analgesic properties in certain types of pain which are not as yet understood. It certainly appears that when analgesia does result from the use of tricyclics, the effect is apparent after 3–4 days, as compared with an antidepressive effect, which can take up to three weeks to develop.

The tricyclic antidepressants all exhibit an anticholinergic effect, and this can be unacceptable in some patients. The resulting dry mouth is a particularly common problem which patients find unpleasant, and this does not generally abate with prolonged use of the drug. Patients who have prostatic enlargement may be precipitated into acute urinary retention, which precludes the use of these drugs in some individuals. Cardiac arrhythmias are occasionally reported, as are visual disturbances, constipation, postural hypotension, and the precipitation of narrow single glaucoma. The pre-existence of some of these conditions may be relative contraindications to the use of these drugs, or at least indicate that extreme caution should be exercised. However, the most usual side effect is sedation and, if daytime somnolence persists, the drugs are unacceptable, particularly in the elderly, or at the other extreme, the active working person, especially if driving is an essential part of their life. There is variation between drugs of this group in the degree of sedation produced, and it is sometimes possible to change to a more suitable drug for an individual. It is also advisable to prescribe the whole dose of tricyclic to be taken an hour or two before the patient's bedtime, so that most of the sedation will occur nocturnally. Of course, in many chronic pain patients this effect may be a bonus, and a positive indication for choosing a tricyclic. Insomnia is a common problem in many patients with chronic pain (although perhaps less so with neurogenic pain) and the sedation and increased relaxation which results from night-time consumption of antidepressants can improve the patient's sense of well-being and improve the results of other pain management procedures. Patients with chronic pain do not generally do well when benzodiazepines are prescribed as tranquillizers for night sedation, and a tricyclic antidepressant will often provide more acceptable anxiolysis and night sedation than diazepam. Although antidepressants are not dependence-forming in the same way as benzodiazepines, withdrawal after prolonged use should be phased, as convulsions can occur after sudden withdrawal of medication.

Some patients will notice an increase in weight when taking tricyclics. This may be a major problem to those already suffering from excess body weight, especially if they have been advised to lose weight in order to help in the treatment of chronic back pain. Fluvoxamine is one of the few antidepressants which do not cause weight gain, and yet may be of benefit in chronic pain as a result of its serotonin re-uptake inhibition. The evidence for its usefulness in this context is at present scanty.

The tetracyclic antidepressants, like mianserin, appear at first to be an attractive alternative to the tricyclics, with less toxicity and fewer side-effects such as the anticholinergic effects. However, these drugs do not appear to possess the analgesic properties that the tricyclic group exhibit. Monoamine oxidase inhibitors appear to have analgesic properties and can sometimes be effective where tricyclics have not. Phenelzine, 15 mg three times daily is perhaps the most widely used agent, but the well-known interactions with diet and other drug therapy make this group of drugs unpopular in Pain Clinics. The conditions in which they have been reported as being effective are atypical facial pain, and thalamic pain.

All of the tricyclic antidepressants have been used in the management of chronic pain. There are differences in their properties, both as far as their mode of action is concerned and in their side-effect profile. For example, clomipramine is thought to have a more serotoninergic activity, whilst maprotiline is more catecholaminergic. Therefore, if a patient fails to respond to one particular drug or if the drug is poorly tolerated, then it is worth trying other drugs in this group. The practitioner will usually become familiar with a small number of the tricyclics, representing different properties.

The archetypal tricyclic antidepressants, amitriptyline, is usually the first choice from this group of drugs when used in chronic pain. It has a mixed action, including catecholaminergic and serotonergic. It also has quite strong anticholinergic effects, and is therefore one of the most toxic tricyclics, and frequently poorly tolerated. However, perhaps because of its mixed actions, it also seems to be one of the most effective drugs in chronic pain, and is therefore widely prescribed. Tolerance can be greatly improved by prescribing very small doses initially, and then increasing over a period of 2–3 weeks. In the elderly it is best to start at 10 mg at night, although younger patients will often tolerate a starting dose of 25 mg at night. Patients should be warned of the morning 'hangover' that may occur initially, and advised if possible to start treatment at a weekend when they do not have to work on the following morning. This sedative effect, while it continues to be useful at night, will generally diminish after a few days and, if the initial dose is tolerated, it can be increased after a week. An analgesic effect may be produced with a dose as low as 10–25 mg, but some patients with chronic pain, especially when there is a marked dysaesthesia from a neurogenic cause, or in some types of idiopathic pain, may require doses of 150 mg at night with further smaller doses in the daytime. Individual tolerance to this drug, especially to the higher doses, varies widely and is not obviously predictable. There is an additive sedative effect with alcohol, and patients should be warned of this. It is probably unreasonable to ask patients to comply with treatment for long periods and totally abstain from any alcohol, as this may further increase their social isolation and reduce their enjoyment of life. Once a patient has stabilized on a dose of amitriptyline, it is probably acceptable to allow them a moderate amount of alcohol, provided that they are aware of the consequences and are warned that they should not try to drink as much as they would normally consider reasonable. Warnings about driving should also be emphasized.

In patients who are unable to tolerate amitriptyline, dothiepin is often an acceptable alternative. It seems to produce much less daytime sedation, dry mouth, or dizziness whilst providing good night relaxation and a valuable degree of anxiolysis. Doses of 25–75 mg at night are commonly prescribed. Clomipramine in a dose of 10–30 mg is effective in some patients and is often well tolerated. It is perhaps less sedative than amitriptyline and may be indicated in chronic pain states that are accompanied by marked behavioural abnormalities.

Other psychotropic drugs

The phenothiazines have frequently been used in the management of chronic pain, but their role is poorly defined. Analgesic properties have been reported, but so have antalgesic properties; it seems that some of these drugs may have a biphasic action, where low doses have an antalgesic action and higher doses are analgesic. The group appears to cover a wide range of claimed therapeutic activities, some being adrenolytic, and others having a direct local analgesic activity on nerve fibres. Drugs in this group which have been reported as having particularly useful analgesic activity include methotrimeprazine, fluphenazine, perphenazine, pericyazine, and chlorpromazine. Generally, the sedative effects predominate but in certain organically-produced neurogenic pain syndromes, such as thalamic pain, they seem to have valuable analgesic properties unrelated to tranquillization. Other 'organic' pain syndromes reportedly improved with phenothiazines are postherpetic neuralgia, causalgia, and back pain associated with nerve damage. They also have a place in the management of idiopathic pains such as atypical facial pain, but in none of these conditions is the role of phenothiazines clearly defined or demonstrated in a controlled way. It is possible that phenothiazones exert their analgesic activity by a central blockade of dopamine receptors, and unfortunately they are therefore associated with the major problems of extrapyramidal side-effects, resulting occasionally in tardive dyskinesia or Parkinsonism in susceptible subjects or with long-term use. They may also be associated with unpleasant subjective experiences, postural hypotension, and weight gain.

The way in which phenothiazines have most often been used in chronic pain has been in combination with tricyclic antidepressants. Useful combinations which are marketed as combined preparations are nortriptyline with fluphenazine, and amitriptyline with perphenazine (Clarke 1981). Unfortunately, fixed-ratio combined preparations all have the disadvantage of not always containing the correct ratio of constituents for an individual patient, but occasionally they are useful when combined therapy is planned, in order to increase patient compliance by reducing the number of tablets consumed.

Another commonly used psychotropic drug in chronic pain is haloperidol. Again this is a dopamine antagonist and, although suggestions of intrinsic analgesic activity may or may not be true, it does seem to potentiate the action of opioid analgesics whilst providing useful anti-emetic effects. Sedation and extrapyramidal effects may present problems. Pimozide is potent psychotropic which has been suggested as useful in managing pain, but the role of this drug is even less well determined.

Clinical syndromes of neurogenic pain

Sympathetically maintained pain

The condition of causalgia has long been recognized as a cause of long-term pain and disability, and yet it frequently follows a relatively trivial injury. Following trauma, usually to a limb, the normal healing process is followed by the development of a painful sensation of burning and hypersensitivity in the extremities. Sometimes this progresses to a full sympathetic dystrophy, or Sudek's atrophy, with dystrophic changes in the skin and its appendages, increased sweating, muscle wasting, and even decreased density of the distal skeleton in the affected limb. The hand or foot are often discoloured, and the patient often complains of changes in appearance of the extremity, at some times being a dusky purple colour, and at others pale or flushed and red. These changes sometimes accompany temperature changes, and cold weather is generally more uncomfortable for the patient. The hand or foot may become swollen and the skin attains a smooth, almost translucent quality. Poor hair growth and a sweaty palm are also indicative of a sympathetic dysfunction. The limb is generally exquisitely sensitive to minimal tactile sensation and the wearing of clothing may become a major source of distress.

The development of a sympathetic dystrophy can occur after major trauma to a limb, such as a crush or degloving injury, but may equally develop following localized nerve trauma or even a closed injury such as a Colles' fracture. It is believed that one of the consequences of peripheral nerve damage is an increase in sympathetic outflow to that segment, but it is more difficult to explain the syndrome that develops after minor or closed injuries.

Development of the condition may be arrested by providing sympathetic blockade to the damaged region shortly after injury, but owing to the unpredictable development of the condition, hard evidence for the benefits of such management is hard to produce. Perhaps more easily demonstrable is the fact that if a sympathetic dystrophy is treated vigorously during the early stages of its development, there is a greater expectation of success than when the condition has been firmly established for a long time.

Apart from a full sympathetic dystrophy, as occurs in a limb, it is not uncommon to see other chronic pains which have an element of sympathetic dysfunction. Some painful scars which are hypersensitive often exhibit changes in skin colour and texture which may represent a sympathetic over-activity. This can sometimes be seen in chronic back pain, especially following surgery or trauma, when the neurogenic component of pain appears to greatly exceed any probable ongoing tissue damage. Such patients often complain of cold, painful legs and may benefit considerably from sympathetic blockade.

Sympathetic blockade can be effected either by direct local anaesthetic (or neurolytic) attack on the regional ganglia, or pharmacologically, which is most effectively achieved by the regional intravenous technique described by Hannington-Kiff (1974).

For sympathetically maintained pain in the lower limb, a lumbar sympathetic block is performed under radiological control. The technique is described in more detail elsewhere in this book. It is advisable to perform a block using only local anaesthetic initially, and then assess the results. Often a single local anaesthetic block will provide relief for several weeks, and the effects of repeated blocks are often cumulative. For this reason it is recommended that a series of blocks with local anaesthetic is preferable to a neurolytic block with phenol when treating sympathetic dystrophy. Unpleasant sequelae, such as chemical neuritis, is avoided—and the results of neurolytic block are often disappointing when block with local anaesthetic has produced good short term results. For this type of lumbar sympathetic block, 8–10 ml of 0.25 per cent plain bupivacaine injected at the level of L3 on one or both sides will produce a good sympathetic block, and most otherwise fit patients can be managed as day cases, provided they are suitably accompanied and are warned of the possibility of short-term postural hypotension.

For the upper limb, a stellate ganglion block can produce good sympathetic blockade of the arm and hand. Local anaesthetic is used and, if effective, the block should be repeated once or twice a week over a period of several weeks in order to produce maximal benefit in relieving the pain of a sympathetic dystrophy.

The patient should lie prone with a pillow under the occiput so that the head is gently extended on the neck. Full resuscitation facilities should be available, and intravenous access may be advisable. It is advantageous to use a needle with a shallow bevel attached to a short, fine bore cannula so that the operator may position the needle and then hold it immobile in position, while an assistant injects the anaesthetic solution. In this way, better control of the injection is achieved. A small skin wheal is raised with local anaesthetic at the level of the sixth cervical vertebra approximately at the level of the cricoid cartilage, and immediately lateral to the trachea. The shallow bevel needle is then inserted at this point between the cartilage and the carotid sheath, in a posterior direction at right angles to the skin. The needle is advanced until the transverse process of the corresponding vertebra is contacted, and then withdrawn a couple of millimetres. The tip of the needle should then be in the correct tissue plane with the sympathetic ganglia, known collectively as the stellate ganglion. The anatomy of this region is shown in Fig. 7.1. The assistant should attempt gentle aspiration with the syringe attached to the narrow tubing on the end of the needle, and if this is negative may proceed with injection. There should be little resistance to injection; if this is not the case, the needle should be repositioned. Similarly, if blood is aspirated, the needle should be repositioned. If, in the rare situation of the needle having penetrated a cuff of dura, CSF is aspirated, the procedure should be temporarily abandoned. After about 1 ml of anaesthetic solution has been injected, the assistant should pause for a few seconds while the patient is observed. We have found this to be a wise precaution, as it is possible for the tip of the needle to penetrate an artery without blood being aspirated. Injection of a small amount of anaesthetic will then result in almost immediate cerebral effects. If the injection proceeds uneventfully, a volume of 10–20 ml of 1% lignocaine are injected. The lower volume is commonly used, although it is possible that 20 mls is required for an adequate block to reach T1. After injection the patient is warned of the effects of the injection, such as temporarily blurred vision (important if they had planned to drive home) and the not infrequent involvement of the laryngeal nerve supply,

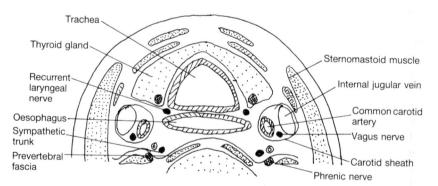

Fig. 7.1 Anatomical relationships in transverse section of the neck.

resulting in temporary hoarseness. If the latter occurs, it is best to advise the patient to avoid drinking until the sensation has worn off.

Although it is possible to produce some effect on sympathetically maintained pain by the systemic administration of adrenolytic drugs, such as labetalol, it was demonstrated by Hannington-Kiff (1974) that good results could be achieved by regional intravenous administration of drugs which block noradrenalin re-uptake at nerve endings. Guanethidine is the agent most commonly used for this technique, but other drugs such as reserpine have been used. Guanethidine initially releases noradrenalin from nerve endings, and then prevents re-uptake. Therefore, the initial effect may be an exacerbation of pain, and this is in addition to the pain produced by the torniquet required for the procedure.

A tourniquet is applied to the upper part of the limb to be blocked; a normal sphygmomanometer cuff is quite suitable, providing it does not leak. The cuff should be applied over a soft layer of protective padding. An indwelling needle such as a 'butterfly' is inserted into a vein on the dorsum of the hand or foot, and a similar needle should be inserted in the contralateral limb to enable any other drugs to be administered in an emergency during the application of the tourniquet on the treated limb. The limb is then partially exsanguinated. This does not need to be as complete as that achieved with an Esmarch bandage and a short period of elevation of the limb is usually enough to empty the veins sufficiently. The cuff is inflated to at least 50 mm Hg above the patient's systolic pressure for an arm, and 100 mm Hg above systolic pressure for a leg, and the limb placed in a position which can be comfortably maintained by the patient for about 20 minutes.

As the procedure is painful, both because of the ischaemia and the initial pain resulting from guanethidine-stimulated noradrenalin release, a local anaesthetic solution can be mixed with the guanethidine. It has been suggested that an alpha-blocking agent be given a few minutes before the guanethidine, but local anaesthetic mixed with it is probably as effective. A dose of 10–30 mg of guanethidine should be sufficient, depending on the size of the limb to be blocked and the build of the patient. For an arm, the guanethidine can be diluted with normal saline and 1 per cent lignocaine in equal volumes to a total of 20–30 ml, again depending on the size of the arm. For a leg, 30–50 ml of the diluted guanethidine will be required. Usually after an initial period of stinging, the patient will become more comfortable, although with time the ischaemia becomes painful. It is probably safe to release the tourniquet, if necessary, after about 5 minutes but, if possible, it should be maintained for 20 minutes, by which time the guanethidine should have adequately spread through the tissues and become relatively fixed. On releasing the tourniquet, the blood pressure should be monitored for a short period, although it is uncommon for hypotension to occur. In fact, a transient rise in blood pressure is occasionally seen as a result of the released noradrenalin from the nerve endings.

The block should ideally be repeated every 4–5 days, but at least once a week, and usually for a minimum of four weeks although some patients may need treatment for longer periods. The guanethidine accumulates in the nerve endings and reduces their ability to respond to autonomic stimulation, so that the duration of relief provided by the block generally becomes extended with successive blocks.

Perhaps surprisingly, some types of centrally-generated neurogenic pain may respond to sympathetic blockade, and this is worth considering in pain which is experienced peripherally as a result of central damage, such as in some cases of thalamic syndrome (Loh *et al.* 1981).

Sympathetically mediated pain can sometimes be ameliorated by systemically administered drugs, which have an effect on the autonomic nervous system. Alpha and mixed alpha/beta receptor blockers, especially labetalol, may be worth a try in this context. The antiserotoninergic agent, ketanserin, has been used to treat sympathetic dystrophies (Hanna and Peat 1989). The drug may be administered intravenously to the affected limb and, if a useful response is achieved, the drug may be continued orally. Although this drug, or more likely a further generation or serotonin blockers, shows promise in the management of sympathetically maintained pains, their use appears to be limited at present by a high incidence of unpleasant side-effects.

Although pain from arterial insufficiency is not neurogenic pain in the generally accepted sense, it is appropriate to mention it here as it is the other main indication for the use of sympathetic nerve blocks. Patients are frequently referred to Pain Clinics for management of pain in the lower limb or ischaemic skin lesions resulting from peripheral vascular disease. These patients can often be relieved of rest pain, and skin perfusion improved by the use of a 'chemical lumbar sympathectomy'. This procedure is usually done as a neurolytic block, using aqueous phenol. The details of this procedure are explained elsewhere, but a single injection of 5 ml of phenol at one level, or two injections of 3 ml each at two different lumbar levels, are usually adequate. Larger volumes increase the risk of unwanted sensory and motor nerve damage, and if the blocks are accurately positioned under radiological control, the smaller volumes are effective. The procedure is often used in an attempt to avoid, or delay, reconstructive surgery and in patients whose condition is considered unsuitable for surgery. It is certainly less traumatic for the patient than either arterial reconstruction or surgical sympathectomy, and the results should be as good as the latter. However, it should be remembered that sympathetic blockade is not generally helpful for intermittent claudication, and may even aggravate this condition by diverting blood flow away from muscle to skin.

Patients with peripheral vascular disease who eventually have an amputation of the affected limb may still benefit from a sympathetic block. There is a possibility that if a sympathetic block or lumbar epidural block is performed prior to amputation, then there is a lower incidence of post-amputation stump and phantom pain. This is an example of the proposal by Bach *et al.* (1988) that the development of pain is reduced if analgesia is provided *before* the onset of the painful 'event'.

Post-herpetic neuralgia

Post-herpetic neuralgia is probably the most common type of neurogenic pain seen in the Pain Clinic, perhaps making up more than 10 per cent of all patients seen. Herpes zoster effects about 125 persons out of every 100 000 of the population per year. In the older patients (over 60 years old), about 50 per cent will go on to develop post-herpetic neuralgia. Although the disease affects younger adults, it is rare to see prolonged pain in those under 40 years old. Although it is believed that herpes zoster is a result of reactivation of a dormant varicella virus, the mechanism of this is not understood. Whereas the condition frequently occurs in those who may be immunocompromised, following chemo- or radiotherapy, or even trauma and surgery, it can equally occur in patients with no obvious predisposing factors. The condition most commonly affects the branches of the trigeminal nerve, and the thoracic and then lumbar segmental nerves, with the chest and abdominal walls being affected. However, the limbs are not uncommon sites, and it is even seen occasionally in the perineum and genitalia.

A necrotizing lesion occurs in the dorsal root ganglion and the dorsal horn of the spinal cord, and it is thought that the virus is transported axonally to the nerve ending where the rash appears. There are widespread degenerative changes in the peripheral nerve. Noordenbos (1959) suggested that pain was due to a destruction of large nerve fibres in relation to a preservation of the smaller fibres, and thus loss of inhibitory activity of the larger fibres, but pathological evidence of this is inadequate. Other suggestions involve a proposed damage to the autonomic ganglia, ectopic impulse generation from partially damaged nociceptors, or intact but abnormally reacting primary nociceptors.

Treatment of the acute phase of herpes zoster infection may influence the development of post-herpetic neuralgia, but controlled evidence is lacking. Recent evidence suggests that treatment of the infection with acyclovir may reduce the incidence of post-herpetic neuralgia (Morton and Thompson 1989), but this is certainly not a universal effect. Unfortunately, the cost of treatment has perhaps reduced the application of this means of reducing the development of a prolonged and incapacitating pain syndrome. Anecdotally, there is a suggestion that the adequate use of potent analgesics during the acute phase can reduce the development of chronic pain, but again hard evidence is difficult to obtain, and certainly once the chronic situation has developed, analgesic drugs are largely ineffective. It has been proposed that the administration of corticosteroids during the acute phase may reduce the incidence of chronic pain. There is also evidence that the administration of sympathetic nerve blocks during the acute phase not only relieve the acute pain, but reduce the incidence of post-herpetic neuralgia (Colding 1973). Again, controlled evidence is lacking, but such procedures do seem to be helpful.

About 10 per cent of patients will develop post-herpetic neuralgia and this seems to be more common when the ophthalmic nerve is affected than for other nerves. The pain characteristically has three components: a constant deep aching pain, sharp lancinating pains, and severe dysaesthesia provoked by light tactile stimulation or allodynia. There may be an area of sensory deficit or, rarely, motor loss. The diagnosis is usually straightforward.

Once post-herpetic neuralgia has become established, treatment is rarely entirely satisfactory but, although the condition may persist for many years, there is a substantial number of patients in whom the symptoms can be ameliorated. The mainstay of treatment has long been established as the use of tricyclic antidepressants. Amitriptyline has been shown to have useful effects in post-herpetic neuralgia (Woodforde et al. 1965) and is still the most widely used of this group of drugs. Intolerance may be a problem, especially in the elderly, but if a starting dose of 10 mg at night is prescribed for the elderly, and the dose is then gradually increased, the drug is more acceptable. Sometimes the beneficial result can be obtained with as little as 10 mg, but often higher doses are required. If this drug is not tolerated, then dothiepin or nortriptyline are possible alternatives. Concurrent administration of pyridostigmine, a cholinesterase inhibitor, in a dose of 30 mg three times a day, may enhance the pain relief provided by amitriptyline while lessening its unpleasant anticholinergic properties, such as dryness of the mouth.

Tricyclic antidepressants are often given in combination with a phenothiazine, the most popular being fluphenazine and perphenazine. The former is available in combination with nortriptyline, and the latter with amitriptyline. However, there are the usual disadvantages of fixed combination drugs, and an adequate dose of tricyclic may be accompanied by an excessive dose of phenothiazine. Many studies suggest an advantage of such combinations (Weis et al. 1982; Taub 1973), but well controlled studies are lacking. A wide range of psychotropic drugs have been suggested in this condition, especially chlorprothixene (no longer available) and pimozide. Occasional success may be seen with these drugs, but controlled studies are not available, and frequently the side effects preclude their use.

Although not universally accepted, there is a general impression that tricyclic antidepressants, with or without phenothiazines, are one of the most effective means of providing symptomatic relief in postherpetic neuralgia. As mentioned above, the pain in this condition may have several components, and the place of the tricyclics seems to be mainly in relieving the dysaesthesic component of the pain. Although some patients are not troubled by their pain during sleep (a common characteristic of neurogenic pain), many find that the pain is worse on retiring, thus delaying sleep, and of course the sedative effect of antidepressants is often a bonus in this situation.

One of the common components of herpetic pain is the lancinating pain which, although intermittent, may be extremely distressing. Patients with this type of pain should be given an anticonvulsant drug. Carbamazepine or sodium valproate are usually the first choice, followed by phenytoin or clonazepam. The combination of amitriptyline with valproate appears to be particularly valuable (Raftery 1979). The combination appears to be more effective than either alone although, again, large controlled trials are not yet available.

Although it may be best to treat acute herpes zoster with nerve blocks in an effort to prevent the development of chronic neuralgia, selective nerve blocks, either peripherally or regionally, may provide prolonged relief (Russel et al. 1957) even when postherpetic neuralgia has been present for many years. Local anaesthetic both with and without the addition of steroid has been used in this way, but the benefits of the addition of steroid are not confirmed. All types of nerve block have at some time been advocated. Epidural and sympathetic blocks have been widely used, but there is no real evidence that they are any more effective in the chronic case than para-vertebral or peripheral blocks. Even subcutaneous infiltration of the affected dermatome with local anaesthetic may provide prolonged relief, and more recently the use of topical applications of local anaesthetic cream in the form of a Eutectic Mixture of Local Anaesthetics (EMLA) (Stow et al. 1989) has been advocated. Although there may be no long-term influence on the condition by the use of nerve blocks, the relief provided will frequently long outlast the action of the local anaesthetic, and often eases the condition for many weeks or months. It has been proposed that the patients who benefit most from local anaesthetic treatment are those in whom there is prominent allodynia. Rowbotham and Fields (1989) suggest that in these patients there is excessive C-nociceptor activity and nociceptor sensitization, activating adjacent receptive fields. Local anaesthesia may block this sensitization. Neurolytic blocks are rarely effective, although success has been claimed with cryogenic nerve blocks. As with most neurogenic pain, it appears that maintenance of some afferent activity is probably important in any management of the pain, and pleas by patients to 'cut the nerve' should be resisted.

In practice, it is worth trying a local anaesthetic block in any patient presenting with post-herpetic neuralgia. In the ophthalmic distribution, blockade of the supra- and infraorbital nerves can be combined with a limited subcutaneous infiltration, and in the thoracic and lumbar regions, para-vertebral blockade can be combined if necessary with local infiltration. For those patients who experience relief outlasting the local anaesthesia, it is worth repeating the blocks at regular intervals. In some, a prolonged effect can be built up which may provide ease for many months. Even in those patients who only experience a couple of weeks of relief, this may be the only method of giving an extremely distressed patient some respite.

Analgesic drugs are often consumed in large quantities by patients with postherpetic neuralgia, commonly without much benefit. The use of potent analgesics in this condition has been widely condemned (Bowsher 1987). There is no doubt that this group of drugs are generally ineffective in this type of pain, and excessive consumption merely produces more unpleasant side-effects. It is normally advisable to change the patient's treatment to drugs such as antidepressants and anticonvulsants. However, it has been mentioned that the pain of postherpetic neuralgia is generally a compound of several different types of pain, and these different components respond individually to different therapies. It also appears that there may be a range of pathological changes producing these varied symptoms. Perhaps because of these factors, there may be a small number of patients who find that the deep aching component of their pain is eased with opioids, although the lancinat-ing pains and allodynia are not influenced. This may account for those few patients who claim that coproxamol is the only agent that gives them any relief, whatever the doctor's theoretical objections to this idea. The response to opioids should be carefully and objectively assessed before any further such drugs are prescribed.

Many surgical procedures have been tried for the relief of postherpetic neuralgia. Most peripheral destructive procedures are at best ineffective, and at worst may aggravate the distress in the long term. Central stimulation techniques may have some promise, but the only procedure that has so far produced results claiming to be better than conservative measures is that of dorsal root entry zone lesion (DREZ lesion) (Friedman and Nashold Jr 1984). Of interest is the possible chemical destruction of nociceptive cells in the dorsal horn using the axonal transport of vinca alkaloids, following transdermal application by means of iontopheresis (Rossano et al. 1989). Work has also been done on the effects of C-fibre degeneration following the application of capsaicin to affected areas of skin. Both of these techniques are in their infancy, but may hold some hope for future development.

Stimulatory techniques are often used for postherpetic neuralgia. Acupuncture has been claimed to be effective in some cases, but is frequently disappointing. Transcutaneous electrical nerve stimulation (TENS) is also helpful in a few refractory cases (Nathan and Wall 1974) although, again, results are generally poor. It is important to remember that if TENS is considered as a form of treatment, there must be adequate sensory function remaining since TENS applied to anaesthetic skin is inevitably ineffective.

Other proposed treatments include ice packs and sprays, vibration, and ultrasound. All of these have their adherents. The application of topical anti-inflammatory agents such as benzydamine (Coniam and Hunton 1988) appears to provide some symptomatic relief, especially from the allodynia, in some patients. The mechanism of this is unknown, and it is difficult to produce good evidence in controlled studies that such treatments are much better than placebo.

Central pain

This term can be used to describe conditions of chronic pain which result from damage to the CNS although, as has been suggested elsewhere in this chapter, many conditions of intractable neurogenic pain may have a central component. However, the term is usually applied to those pain syndromes which are believed to arise as a direct result of ischaemic, haemorrhagic, traumatic, neoplastic, or congenital lesions arising primarily within the brain or spinal cord. The original descriptions were applied to lesions of the thalamus and, although this is not the only cause of such syndromes, the term 'thalamic syndrome' is often used generically when the condition develops following a cerebral haemorrhagic episode. The condition may present with a variety of neurological signs and symptoms, but when pain is present it usually occurs as an intractable spontaneous pain of a dysaesthetic type, often with allodynia, and sometimes accompanied by lancinating pains. There may or may not be a sensory deficit. In many ways, this represents the classical picture of a neurogenic pain. The pain and sensory disturbance from a lesion in the thalamus or from a cortical or subcortical lesion usually affect the contralateral side of the body to the lesion, but are often particularly localized to the face, mouth, hands, and feet.

Similar problems can occur after transection of the spinal cord, with the development of phantom sensations or lancinating and dysaesthetic pains, as well as the reflex spasms occurring in the affected limbs. Degenerative conditions of the spinal cord, such as syringomyelia, multiple sclerosis, subacute combined degeneration, and tabes can also produce centrally generated neurogenic pain. Other examples may follow neuroablative surgical procedures.

The management of central pain is always difficult, and usually unsatisfactory. A wide variety of drugs have been tried, but those used in other types of neurogenic pain are often the most effective. The anticonvulsant drugs are often useful in controlling the lancinating pain, although high doses are often required. Tricyclic antidepressants, especially if combined with a phenothiazine such as fluphenazine, sometimes help to reduce the burning dysaesthetic pain. The effect of tricyclics may be further enhanced in some cases by the addition of L-tryptophan, a precursor of 5-HT, again perhaps enhancing the intrinsic pain inhibitory mechanisms.

Nerve blocks are generally unsatisfactory, although sympathetic blockade can sometimes help to relieve centrally generated pain.

Much interest has been aroused by the use of opioid antagonists in the management of thalamic syndrome. It has been noted that in some instances this type of pain, which has been refractory to other forms of treatment, responds to naloxone treatment. Large doses of 4–10 mg have been infused intravenously, and a number of such patients experience pain relief which may last only a few hours, or sometimes for several weeks. Patients may develop cardiac arrhythmias during rapid administration of such doses of naloxone, and it is advisable to infuse the drug over about half an hour during continual electrocardiogram (ECG) monitoring. Those that show a reasonable response may have the procedure repeated, or transfer to the oral drug, naltrexone, for maintenance (Budd 1987). Studies have been conducted concerning blood flow in the region of the thalamus following naloxone, but the mechanism of action is at yet obscure, and some authorities have recently thrown doubt on the validity of the treatment.

Post-amputation pain

There is a variety of clinical problems that may develop following amputation which have some similarities whether the amputation involved physical loss of a limb (it may also occur following mastectomy to produce a phantom breast) or avulsion of the nerve plexus supplying that limb resulting in total deafferentation. Experience of a sensation of the intact limb being present, which can involve all sensory modalities, including movement, is commonly experienced by amputees. This is the classical phantom limb. The sensation is not necessarily unpleasant, although perhaps this depends on the condition of the limb before loss. Many patients feel embarrassed to admit the presence of phantom sensations, and are relieved when it is explained that this is quite normal. Although the sensations tend to fade gradually with time, often with a feeling of telescoping of the phantom limb so that the limb becomes shorter and the hand or foot is felt more proximally, in many patients the phantom will persist indefinitely.

Unfortunately, in many patients the phantom limb is painful, and this is often described as a crushing or tearing pain, most frequently in the hand or foot. Pain seems to be more common when the limb has been painful prior to loss and, as mentioned earlier, relief of pain pre-operatively may be an important prophylactic measure when the amputation is elective.

Some post-amputation pains appear to arise in the stump. This may be due to localized stump trauma from poorly fitting prostheses, bone spurs, or from neuroma formation. The pain which is believed to arise from neuromas often presents as acutely tender areas, or trigger points on the stump. Tactile stimulation of these points may give rise to stabbing or electric shock-like pains, sometimes accompanied by clonic movements of the stump. This type of pain is neurogenic pain of peripheral origin, whereas phantom pain is presumably of a central deafferentation type. The two conditions frequently coexist.

Deafferentation with a relatively intact limb is most vividly illustrated in the brachial plexus avulsion injury, commonly seen following motor cycle accidents. The paralysed and anaesthetic arm is a focus of severe phantom pain, again most often of a crushing nature. Amputation of the useless arm does not relieve the pain.

The pain of an amputated limb is generally aggravated by the emotional impact and disability of the injury. Problems with adapting to a prosthesis may be a further contributory factor. Successful physical and social rehabilitation, combined with other distracting stimuli, tend to help with management of the pain.

Treatment of post-amputation pain is difficult. Stump pain from neuromata can often be eased by injecting local anaesthetic into the painful neuroma. If this is successful, the procedure can be repeated with phenol or cryotherapy. If the pain is characterized by lancinating shock-like pains then patients often gain some relief from the use of the anticonvulsant drugs.

Phantom pain or the pain following plexus avulsion can sometimes be partially eased with tricyclic antidepressants, often with the addition of a phenothiazine such as perphenazine. Anticonvulsants may need to be added if lancinating pains are also present.

Destructive techniques are rarely satisfactory. Further amputation does not affect the pain, and neither does rhizotomy. Cordotomy and sympathectomy may produce some relief, but the effect is generally temporary and the long-term consequences of cordotomy may be worse. DREZ lesions have been advocated, but results are not very encouraging.

Stimulation techniques may offer hope to the amputee. TENS is often helpful, and has the advantage of being non-invasive. In some patients it is the only therapy which offers any relief, however incomplete. The electrodes can be applied over the stump and/or the remaining nerves. Following brachial plexus avulsion it is obviously pointless to apply TENS to the limb, but occasionally some relief can be produced by using the electrodes over the shoulder where the zone of sensation begins. Spinal cord stimulation has been used with success for post-amputation pain and, where facilities exist, it is worth considering a trial. Deep brain stimulation has also been used with some success.

Acupuncture can provide relief from phantom pain in some patients. As the effect of acupuncture is generally bilateral, it is possible to apply needle stimulation to the contralateral intact limb and provide some relief to the injured side. It has been suggested that this method produces analgesia by stimulating bilaterally the descending inhibitory systems with the spinal cord. This effect may be enhanced by simultaneous use of tricyclic antidepressants and, perhaps, by the administration of tryptophan.

Peripheral neuropathies

There are many causes of peripheral degeneration of the nervous system: congenital, metabolic, toxic, ischaemic, neoplastic, and traumatic. These may result in selective large fibre loss (as in Friedreich's ataxia), small fibre loss (as in diabetic neuropathy), or non-selective loss (as in alcoholic neuropathy). Treatment of these conditions tends to rely on tricyclic antidepressants with anticonvulsants where there is a lancinating component. A controlled trial of nortriptyline combined with fluphenazine in the treatment of painful diabetic neuropathy demonstrated significant relief of pain and paraesthesia (Gomez-Perez *et al.* 1985). Generally, this type of drug therapy seems to be one of the most useful means of dealing with peripheral neuropathies but, in some instances, the additional use of TENS can be helpful.

Cranial nerve neuralgias

Trigeminal neuralgia, or tic douloureux, is generally a relatively straight-forward diagnosis when the classical history is described by a sufferer although facial pain with some characteristics of this condition, whilst not fitting the complete picture, may occur. The condition is characterized by brief bursts of intense electric shock-like pain in the distribution of one of the branches of the trigeminal nerve. A similar, but rarer, condition can affect the glossopharyngeal nerve or the nervus intermedius of the facial nerve. The pain is usually triggered by non-nociceptive tactile or thermal stimulation of the face or oral region on the same side as the pain. In between pain bursts there is little discomfort, and rarely more than minimal sensory disturbance. The pain bursts may come as groups in rapid succession, and the condition is often characterized by periods of spontaneous remission, sometimes lasting months or even years. More than one division of the nerve may be affected, and the commonest to be affected is the mandibular branch, with the ophthalmic branch the least common. Although the condition may affect the same patient on both sides at different times, it is extremely rare for this to occur simultaneously. Pain in the glossopharyngeal nerve may be accompanied by bradycardia and syncopal attacks. The pain of these neuralgias is usually excruciating, and the fact that it is triggered by minimal tactile stimuli may render patients almost frightened to move, talk, or eat during a period of attacks.

The aetiology of trigeminal neuralgia is obscure. It is a fairly common development in multiple sclerosis. Other cases may be due to compression of the nerve at the level of the pons, often by blood vessels or by tumours. It has been suggested that some trigeminal neuralgias may develop as a deafferentation response to extraction of teeth.

The first line of treatment for trigeminal neuralgia is almost always a trial of carbamazepine. Since it was discovered that this drug is usually effective in trigeminal neuralgia, it has almost been regarded as a diagnostic criterion for the condition that a response to it is obtained. It is usual to start with a low dose, such as 100 mg twice daily, and increase gradually every few days, initially up to 200 mg four times daily and, in a few cases it may be necessary to increase up to a total of 1800 mg per day. Many patients will obtain relief at lower levels. Carbamazepine does have unpleasant side-effects, such as dizziness, ataxia, somnolence, and nausea, and for some patients these become intolerable before a therapeutic dose can be obtained. If relief is obtained, but with intolerable side-effects, it may help to stop the medication for 24 hours and then begin at a lower dose again. Bone marrow suppression can rarely occur, so routine haematological checks should be made during the first months of therapy. Approximately 50 per cent of patients can be successfully managed with carbamazepine but, unfortunately, tolerance often develops after prolonged therapy and this may be due to autoinduced metabolism of the drug. As the condition often follows an intermittently relapsing course, treatment could be stopped during a quiescent phase and restarted, often with renewed sensitivity, when symptoms reappear. Alternatively, an alternative anticonvulsant can be tried. This should also be tried if the condition does not respond initially to carbamazepine or if side-effects are intolerable.

Sodium valproate is at least as effective as carbamazepine in trigeminal neuralgia, and is often better tolerated. Again, the dose can be gradually increased from 100 mg two or three times a day, up to a maximum of 1200 mg daily. It is best taken at meal times to reduce gastrointestinal intolerance, but nausea is still occasionally a problem. Other side-effects are of a similar nature to those with carbamazepine, plus the occasional incidence of alopecia. The latter is generally reversible on cessation of therapy. Disturbance of hepatic function can occur during valproate therapy, and liver enzymes should be checked prior to starting and at intervals during the first few months. A rise in blood levels of hepatic enzymes may not necessarily reflect potential liver damage, but if large changes in function are seen, then a change of medication is probably wise.

Phenytoin was the original drug used in this condition, but it is now regarded as a third line drug. CNS and dermatological side-effects are fairly common, and it is usually recommended that plasma levels are monitored. After a starting dose of 100 mg three times daily, plasma levels can be monitored at three weeks, and the dose adjusted to produce a plasma level of 15–25 μg/ml, which is usually adequate to provide relief if the patient is going to respond to this drug. In some resistant cases, combining phenytoin with carbamazepine or sodium valproate may produce a better therapeutic effect than either drug used singly.

Clonazepam represents another class of anticonvulsants which is sometimes useful in neuralgias not responding to other anticonvulsants. The dose should start at 0.5 mg at night, and be increased gradually up to a maximum of 4 mg at night. Many patients cannot tolerate the higher doses, and some cannot tolerate the sedation and dizziness that may result even from the smallest dose.

Various other drugs have been described in the management of this condition, but baclofen is one of the few that may have a place in treating difficult cases. Local anaesthetics have been tried with varying success. Certainly, the use of local anaesthetic gargles may help to provide relief of glossopharyngeal neuralgia. Although the pain is of central origin, it is generally triggered by peripheral stimuli and anaesthetizing the trigger points is sometimes effective in the short term. Cryotherapy is applied to branches of the trigeminal nerve, usually by means of a surgical exposure. Adherents of this technique claim good medium-term results and, in many cases, have been able to repeat the procedure several times. Acupuncture seems to reduce the severity of the condition in a few individuals, stimulating both local and distal points. However, the method is not universally successful, and it is difficult to evaluate the treatment objectively.

In cases which are resistant to drug therapy, or gradually become so, or in those patients who are unable to tolerate adequate doses of medication, surgical intervention may eventually be required. Trial blocks of the trigeminal nerve divisions with local anaesthetic may help in selection of patients suitable for destructive procedures, and can produce welcome short-term relief for those unsuitable for, or awaiting, surgery or permanent nerve block. Neurolytic injection of the gasserian ganglion is widely practised, especially where neurosurgical services are not readily available. The approach to this ganglion is usually that described by Härtel (1912). The target is the foramen ovale, which is located medial to the tubercle of the zygoma. The procedure should be performed with the assistance of radiographic guidance.

A skin wheal is raised over the second upper molar tooth, about 3 cm lateral from the line of the corner of the mouth. A 10 cm 20 gauge needle is then advanced so that, when viewed laterally, it is aimed at the mid-point of the zygomatic arch and, from the front, aimed at the ipsilateral pupil (Fig. 7.2). The needle is advanced upwards, medially and posteriorly until it contacts the infratemporal plate, lateral to the base of the pterygoid process, and anterior to the foramen ovale. A marker is then placed on the needle at

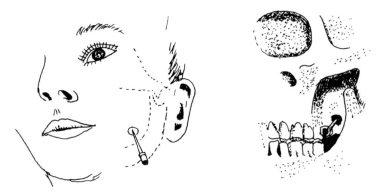

Fig. 7.2 Gasserian ganglion blockade.

1.5 cm from the skin, and the needle withdrawn into the subcutaneous tissues. It is then redirected slightly more posteriorly so that it enters the foramen. The point of the needle is considered to have reached the gasserian ganglion when it lies 1.5 cm deeper than the bone contact. Paraesthesiae in the trigeminal distribution may result as the needle passes through the foramen. If the needle is aimed at the lateral part of the ganglion, the third division is more likely to be blocked, and if in the more medial portion, then the first and second divisions are affected. If neither blood nor CSF can be aspirated, 0.5–1 ml of local anaesthetic can be injected if it is proposed to do a trial block. This will give the patient a chance to experience the sensation of numbness before a decision is made to perform a destructive lesion. Entry of the ganglion itself can be extremely painful, and it may be helpful to briefly use a general anaesthetic during this procedure. Up to 1 ml of absolute alcohol is injected for a neurolytic block, although better results may be produced using phenol in glycerine in 0.1 ml increments until the desired effect is achieved (Jefferson 1963). If the injection is not actually into the ganglion, sensation may return to the upper two divisions within a few days, but is likely to be more effective for the third division of the nerve. The use of phenol mixed with X-ray contrast medium may enhance radiographic localization of the block. It may be that a prolonged effect results from spread to the sensory roots at the back of the ganglion. Glycerol injection has also been successfully used to produce lesions in the trigeminal nerves, and this agent is said to result in less sensory loss, but a poorer success rate and a higher incidence of relapse.

A similar technique can be used to introduce an electrode through the foramen to the sensory roots of the nerve in the subarachnoid space. A radio-frequency thermocoagulation lesion can be produced. It is possible to produce a more selective lesion of the C-fibres by this technique, thus preserving

much normal sensation. Prior to the production of the lesion, the exact localization of the electrode can be determined by sensory testing with a current of
50–100 Hz, and then motor testing with a low frequency current to ensure
adequate distance from the motor fibres. The patient can then be given a
short-acting general anaesthetic while a series of lesions is produced.

Destructive lesions of the trigeminal nerve are associated with a certain
amount of morbidity. Anaesthesia dolorsa is a possible complication of the
procedure. This usually develops gradually after the lesion, and is associated
with a painful burning dysaesthesia and numbness in the area of distribution
of the affected nerve. A reddening and coarseness of the skin may also be
apparent, and it has been suggested that this may represent a severe form of
sympathetic dystrophy. Treatment of anaesthesia dolorosa is generally
unsatisfactory, but sometimes partial relief may be provided by the drugs
otherwise used in neurogenic pain, particularly clonazepam, and perhaps the
combination of amitriptyline and perphenazine. Another complication that
may follow sensory destruction is the development of trophic lesions in the
area of anaesthesia. Occular palsy and corneal anaesthesia are also
occasional complications.

There is a variety of neurosurgical operations that have been used to
manage trigeminal neuralgia, such as medullary tractotomy and trigeminal
rhizotomy, but perhaps the most satisfactory is exploration of the posterior
fossa and decompression of the sensory root. Sometimes, an arterial loop is
found to be compressing the sensory root and a non-absorbable sponge can
be inserted between these structures to relieve the compression. This
technique has been used to relieve neuralgias of the ninth and even the tenth
and eleventh cranial nerves, and some surgeons quote a rate of 85 per cent of
patients relieved of their symptoms. This does, however, constitute a major
neurosurgical procedure and the morbidity needs to be considered; cerebellar infarct, meningitis, occular paresis, facial paralysis, hearing impair-
ment, and trigeminal sensory loss have all been reported. Generally,
however, it is well tolerated and there is a low incidence of sensory loss of
dysaesthesia. It is often the treatment of choice in younger patients, and
certainly when the condition is bilateral. In older patients, and in cases of
failure or recurrence, thermocoagulation or glycerolysis may be preferable.

Glossopharyngeal neuralgia may be treated in the same pharmacological
way as trigeminal neuralgia, or by blockade of the ninth nerve, or by thermocoagulation, or microvascular decompression.

References

Arner, S. and Meyerson, B. A. (1988). Lack of analgesic effect of opioids on neuro-
    pathic and idiopathic forms of pain. *Pain*, **3**, 11–23.

Bach, S., Noreng, M. F., and Tjellden, N. U. (1988). Phantom limb pain in amputees during the first 12 months following limb amputation, after preoperative lumbar epidural blockade. *Pain*, **33**, 297–301.

Bowsher, D. R. (1977). Neurogenic pain. *Journal of Intractable Pain Society*, **5** (1), 23–7.

Budd, K. (1987). Clinical use of opioid antagonists. *Clinical Anaesthesiology*, **1**, 993–1011.

Clarke, I. M. C. (1981). Amitriptyline and perphanazine in chronic pain. *Anaesthesia*, **36**, 210–12.

Colding, A. (1973). Treatment of pain: organisation of a pain clinic. Treatment of acute herpes zoster. *Proceedings of the Royal Society of Medicine*, **66**, 541–3.

Coniam, S. W. and Hunton, J. (1988). A study of benzydamine cream in postherpetic neuralgia. *Research and Clinical Forums*, **10**, 65–7.

Friedman, A. H. and Nashold Jr. J. S. (1984). Dorsal root entry zone lesions for the treatment of postherpetic neuralgia. *Neurosurgery*, **15**, 969–70. **60**, 1258–62.

Gomez-Perez, F. J., Rull, J. A., Dies, H., Rodriguez-Rivera, J. G., Gonzalez-Barranco, J., and Lozano-Castaneda, O. Nortriptyline and fluphenazine in the symptomatic treatment of diabetic neuropathy. A double blind cross over study. *Pain*, **23**, 395–400.

Hanna, M. H. and Peat, S. J. (1989). Ketanserin in reflex sympathetic dystrophy. A double blind placebo controlled cross-over trial. *Pain*, **38**, 145–50.

Hannington-Kiff, J. G. (1974). Intravenous regional sympathetic block with guanethidine. *Lancet*, **1**, 1019–20.

Härtel, F. (1912). *Langenbeck's Archiv Fur Chirurgie*, **100**, 193.

Jefferson, A. (1963). Trigeminal root and ganglion injections using phenol in glycerine for the relief of trigeminal neuralgia. *Journal of Neurology, Neurosurgery, and Psychiatry*, **26**, 345.

Johansson, F., Von Knorring, L., Sedvall, G., and Terenius, L. (1980). Changes in endorphins and 5-hydroxyindoleacetic acid in cerebrospinal fluid as a result of treatment with a serotonin reuptake inhibitor (zimelidine) in chronic pain patients. *Psychiatry Research*, **2**, 167–72.

Lee, R. and Spencer, P. S. J. (1977). Antidepressants and pain: a review of the pharmacological data supporting the use of certain tricyclics in chronic pain. *Journal of International Medical Research*, **5**, 146–56.

Loh, L., Nathan, P. W., and Schott, G. D. (1981). Pain due to lesions of the central nervous system removed by sympathetic block. *British Medical Journal*, **282**, 1026–8.

McQuay, H. J. and Dickenson, A. H. (1990). Implications of nervous system plasticity for pain management. *Anaesthesia*, **45**, 101–2.

McQuay, H. J., Carroll, D., and Moore, R. A. (1988). Postoperative orthopoedic pain—the effect of opiate premedication and local anaesthetic blocks. *Pain*, **33**, 291–5.

Morton, P. and Thompson, A. N. (1989). Oral acyclovir in the treatment of herpes zoster in general practice. *New Zealand Medical Journal*, **102**, 93–5.

Nathan, P. W. and Wall, P. D. (1974). Treatment of postherpetic neuralgia by prolonged electrical stimulation. *British Medical Journal*, **3**, 645–7.

Noordenbos, W. (1959). *Pain*. Elsevier, Amsterdam.

Raftery, H. (1979). The management of postherpetic pain using sodium valproate and amitriptyline. *Irish Medical Journal*, **72**, 399–401.

Rossano, C., De Lucat, L. F., Firetto, V., Fossi, F., Vannini, S., Memeo, R. *et al*. Vinca alkaloids administered by iontophoresis in postherpetic pain: a preliminary report. *The Pain Clinic*, **3**(1), 31–6.

Rowbotham, M. C. and Fields, H. L. (1989. Postherpetic neuralgia: the relation of pain complaint, sensory disturbance, and skin temperature. *Pain*, **39**, 129–44.

Russel, W. R., Espir, M. L. E., and Morganstern, F. S. (1957). Treatment of post-herpetic neuralgia. *Lancet*, **1**, 242–5.

Stow, P. J., Glynn, C. J., and Minor, B. (1989). EMLA cream in the management of postherpetic neuralgia. Efficacy and pharmacokinetic profile. *Pain*, **39**, 301–5.

Swerdlow, M. (1980). The treatment of shooting pain. *Postgraduate Medical Journal*, **56**, 159–61.

Swerdlow, M. (1986). Anticonvulsants in the therapy of neuralgic pain. *The Pain Clinic*, **1**, 9–19.

Taub, A. (1973). Relief of postherpetic neuralgia with psychotropic drugs. *Journal of Neurosurgery*, **39**, 235–9.

Weis, O., Sriwatanakul, K. and Weintraub, M. (1982). Treatment of postherpetic neuralgia and acute herpetic pain with amitriptyline and perphenazine. *South African Medical Journal*, **62**, 274–5.

Woodforde, J. M., Dwyer, B., McEwen, B. W., De Wilde, F. W., Bleasel, K., and Conneley Jr, T. J. (1965). Treatment of postherpetic neuralgia. *Medical Journal of Australia*, **2**, 869–72.

8 Headache

The cause of the vast majority of headaches is unknown; it is even a matter of debate as to whether the pain of headache originates in intracranial or extracranial structures. If a sufferer from headache is referred to a Pain Clinic a logical approach to providing help must start from this point of ignorance.

As headache may be, although very rarely is, a symptom of progressive intracranial disease, the clinician's first task must be a thorough assessment, as in all pain relief work. If the headache is associated with signs of damage to the nervous system, in particular with visual field defects, changes in consciousness, or sensory or motor loss, then the headache may be due to a space-occupying lesion. The patient should be examined to confirm that these symptoms are associated with appropriate physical signs and referred to a neurologist for further management immediately. If the headache is of sudden and recent origin or if it is associated with neck stiffness and, perhaps, pyrexia then it may indicate infection or haemorrhage and once again the patient should be in the hands of a neurologist or neurosurgeon. The management of headache in a Pain Clinic illustrates a major facet of all pain management; that is, that meticulous attention to detail in history taking and examination is the only way to be sure that patients do not get symptomatic treatment for pain when therapy for its cause is essential. It is also the only route that will provide us with the information that will link the symptom of pain to physical findings and increase our understanding of its causes.

In the absence of a known cause or origin for chronic headache, the traditional medical approach has been to classify it. Classification will be of little use to the pain clinician, as the few clues that it does give for clinical management will almost invariably have been tried and found wanting before the patient reaches a Pain Clinic. However, until a better way of thinking about headache is generated by a more thorough understanding of its pathology, it is the best thing that we have.

Throbbing headaches have been assumed to be of vascular origin. They are almost always associated with vascular changes and, in many cases, changes in plasma serotonin levels respond to drugs that affect the vascular system or serotonin receptors. However, the role of serotonin in headache is not understood and, as the blood vessels of the brain have no nociceptive innervation, the pain cannot start in them or their walls. We can only assume that the pain must arise in association with the vascular changes, not directly because of them.

Because these headaches are mainly unilateral they were originally given the name hemicrania but are now commonly known as migraine. Classical migraine is a unilateral headache with photophobia, nausea, and vomiting, often initiated by stress. It has a prodromal phase in which there is a change of mood over the preceding day or so. An aura develops, with visual changes such as a blurring of vision, a scotoma, visual field changes, sensory changes such as paraesthesiae, and motor disturbances which may involve speech. Common migraine is a unilateral headache with nausea and photophobia, but without an aura. In carotidynia the pain is experienced in the structures supplied by the external carotid artery which is itself tender to pressure. In cluster headache the unilateral headache recurs in separate bouts of daily attacks over a period of several months. The attacks are extremely severe, and occur in groups of one to three a day. The pain is accompanied by rhinorrhoea and lachrymation. In chronic paroxysmal hemicrania the pain occurs daily and is felt principally in the ocular, frontal, and temporal areas. The characteristic feature of chronic paroxysmal hemicrania is the complete relief afforded by daily indomethacin.

Virtually all of us have sometimes developed a tension headache in response to stress. It will have been short-lived and responded to simple analgesics. Where such a headache becomes severe and chronic, however, the sufferer may be referred to a Pain Clinic. The diagnosis of tension headache is exclusive. It is not caused by intra- or extracranial pathology, it is only rarely unilateral, it is not usually associated with vomiting, but there is a strong association with anxiety and depression. Pain arising from the temporomandibular joint has been described in Chapter 5 and it is suggested it is one of the myofascial pains. It may be that many other varieties of headache also belong to this category.

Trigeminal neuralgia is a facial pain rather than a headache. It is a lancinating pain and is of such an intensity that it is usually scored at the top of a visual analogue pain scale. It is experienced in the distribution of one, or occasionally two, branches of the trigeminal nerve. The pain is triggered by a stimulus such as touch or cold. Remissions are common. If the pain does not fulfil these diagnostic criteria exactly, then the diagnosis should not be considered.

Trigeminal neuralgia may be a symptom of a small posterior fossa tumour, of a demyelinating disease, or the result of compression by an abnormal artery of the area of the medulla from which the trigemial nerve roots emerge. Changes in myelination of this area in sufferers who do not have multiple sclerosis have been found. Before symptomatic treatment is started, lesions which might be responsible should be sought.

Treatment

It would be unusual for a patient to reach a Pain Clinic suffering from headache who was not receiving conventional mild analgesia. For most migraine sufferers a mild analgesic, together with an anti-emetic when nausea is a prominent component of the migraine, will control the pain of attacks. Other migraine therapies are based on unproven theories about the cause of migraine. There are vascular changes in migraine starting with a vaso-constriction followed by a vasodilation. Migraine and cluster headache attacks can be precipitated by nitroglycerine, a vasodilator. Ergotamine, which is a vasoconstrictor, will relieve the pain of migraine. However, the evidence that the vascular changes occurring in migraine are the cause of the pain is poor. Ergotamine is a vasoconstrictor and provides pain relief in migraine but it is also an analgesic and this may be responsible for the relief that it provides. Migraine pain is not only intracranial; there are pronounced areas of muscle tenderness during migraine attacks, particularly in the temporalis muscle but also in the muscles of the neck, so the pain is not exclusively 'vascular' (Jensen *et al*. 1988). There are changes in serotonin levels in migraine and pizotifen, a drug that is both an antihistamine and antiserotoninergic, is an effective migraine prophylactic. Beta-blockers and antidepressants are also used in the prophylaxis of migraine. Thus, the picture of migraine—its causation, treatment, and prophylaxis—is rather confused.

Where all conventional methods of helping the migraine sufferer have failed, the pain clinician should first survey the methods so far used to confirm that they were given a proper trial. Pain Clinic methods such as trigger point injection of the tender pericranial areas, or acupuncture needl-ing of them, can provide prophylaxis and pain relief that is just as effective as any conventional medication, and safer. Where a patient has previously been wrongly treated, for example when techniques directed at other headaches (such as temporomandibular joint dysfunction or atypical facial pain) have failed, it is worth bearing in mind that there may be a migraine origin to the pain and using pizotifen prophylactically for a time.

In tension headache, once again, the clinician will need to be certain that conventional mild analgesia has been effectively tried. In all pain manage-ment where the aim is to use analgesic drugs to control pain, they must be given at such a dose and frequency that the pain is not experienced. If this level of analgesia can be achieved and the patient is symptom free, they will think that they are cured. If not, no matter how effective is the temporary analgesia, the patient will remain a sufferer. Sensitivity to the efficacy and to the side-effects of the large variety of mild analgesics varies tremendously. The only way in which to see if effective pain relief can be provided by mild analgesics is to use a wide variety of them up to the maximum safe or acceptable dose. Paracetamol is the safest and best tolerated of the available mild analgesics and should be tried in a dosage of up to 1 g 6-hourly before other more dangerous drugs. In tension headache, as in migraine, there is pericranial muscle tenderness and acupuncture needling or trigger point injection with local anaesthetic should also be given a trial.

If the pain is almost entirely occipital, it is possible that it is the result of occipital nerve damage. This will be accompanied by sensory changes, in the distribution of this nerve. Blockade of the nerve with local anaesthetic may confirm the diagnosis, and it may be possible to relieve the pain for longer periods by injecting it with depot steroids.

The pain of trigeminal neuralgia can be relieved by anticonvulsant drugs. Carbamazepine dosage should be rapidly increased until the pain is controlled. Although carbamazepine has traditionally been the first choice of treatment in trigeminal neuralgia, there is no particular logic in using it rather than other anticonvulsants; it has no special merits, and in many patients sodium valproate will provide complete pain relief with fewer side-effects. When one anticonvulsant becomes ineffective then the other should be tried. Their effects may be reinforced by adding clonazepam. However, as this drug is a benzodiazepine it should be introduced with circumspection because of the problems of dependence that usually develop with this drug.

Trigeminal neuralgia may not respond to medication, or the patient may become insensitive to it. When this happens, there is a need to resort to nerve blockade and destructive techniques. As our understanding of the mechanism of trigeminal neuralgia is at best a matter of dispute, destroying parts of the nervous system to relieve it, even if this is effective, must be poor medicine. There is no doubt that a technique as simple as infiltration of the trigger area may produce a prolonged remission early in the disease. Nerve ablation, either chemical or by avulsion peripherally, will produce longer remissions, and the more proximal the destructive lesion is, the more profound and prolonged the response (and, of course, the greater the subsequent disability). The ultimate percutaneous destructive lesion is damage to, or destruction of, the trigeminal ganglion.

As has already been mentioned in Chapter 7, a variety of techniques has been used to damage the ganglion (see pp. 124–6). It is becoming clear that the extent of the damage is not related to the success of the treatment. Alcohol was used in the earlier techniques and the result was complete analgesia, but this often led to corneal damage after block of the ophthalmic division and could result in damage to other cranial nerves when the alcohol went astray. Phenol in glycerine is more controllable and less destructive, and it is usually possible to preserve light touch with loss of pin prick and relief of pain. Radio-frequency lesioning has made it possible to make lesions more precisely defined. A radio-frequency lesion maker has three functions. It can operate at frequencies that will stimulate either sensory or motor nerve, it can measure impedance (which will increase when the needle is in fatty tissue such as the tracts of the spinal cord), or, for lesioning it can heat while measuring tip temperature. The needle can be manœuvred inside the skull until a stimulus shows that it is over the part of the ganglion through which the trigger is passing.

Two newer techniques, the injection of glycerine alone and the inflation of a small balloon over the ganglion to produce a temporary compression, will both relieve trigeminal neuralgia with the minimum of neurological deficit and risk of damaging other cranial nerves. We do not know the cause of trigeminal neuralgia, and until we do we will be unable to understand the reasons why destructive or other lesions to the nerve produce pain relief. At our present level of knowledge, it would seem unjustifiable to damage more nervous tissue than is absolutely essential.

The most consistent and long-lasting results have come from micro-vascular decompression of the trigeminal roots within the posterior fossa (see Chapter 7, p. 126). There is some evidence that certain patients with trigeminal neuralgia have an abnormality of the anatomy of the posterior cerebellar artery, bringing it abnormally close to the trigeminal nucleus in the pons. It is suggested that the pulsation of the artery against the brain stem causes demyelination and thus neuralgia. Placing a small sponge between the artery and the brain stem seems to produce a more prolonged relief than any other technique yet devised. However, this does involve major surgery and there is debate about whether the abnormality of the artery really is linked to the success rate, thus whether this really is the cause of neuralgia. So, the problem is certainly not yet solved.

As our ignorance in this area is almost entire, it is not surprising that the results of therapy are poor. Occasionally, the pain clinician viewing the problem from a different viewpoint to that of the neurologist or other specialist may realize that the wrong route has previously been followed in the management of a patient's headache. The most important contribution that Pain Clinic methods will make is the recognition that stress and other triggers play a part in the initiation and continuation of most chronic head-aches. Stress management techniques such as cognitive therapy and training in relaxation may have more to offer to prevent headache than prophylactic medication and invasive surgery.

Reference

Jensen, K., Tuxen, C., and Olesen, J. (1988). Pericranial muscle tenderness and pressure pain threshold in the temporal region during common migraine. *Pain*, **35**, 65–70.

9 The atypical pains

This term is generally taken to include a collection of pain syndromes which do not correspond to any known pathology. Although this could, perhaps, include a large proportion of chronic pain syndromes, the term is most commonly used to describe a syndrome of facial pain not typical of trigeminal neuralgia and, sometimes, a similar syndrome presenting with a complaint of pain in the perineum or genitals. The term 'atypical pain' is most unsatisfactory, as it really means nothing to any observer except to the person who originally coined the phrase for pain which seemed to be an atypical variant of some other recognized condition. To be preferred is the expression 'idiopathic pain' with its implication of a disorder of pain perception, or 'pain of unknown aetiology'. The latter suffers from the disadvantage of being too wide a description. Since 'atypical facial pain' is such a widely used diagnostic term, it is the one that will be used in this chapter and is, perhaps, better than covering all pain of unknown aetiology by the description 'psychogenic pain'. This description is commonly used, but it makes assumptions about the cause of the pain which cannot always be justified, except by a process of exclusion.

Mention will also be made in this chapter of related syndromes, such as glossodynia, and other common conditions without obvious pathology, such as some types of headache, abdominal pain, whole limb pain, and pelvic pain. The description of some of these as idiopathic pain may be controversial, and their aetiology may eventually be determined. The common features of many of these pain syndromes are that they often do not fit with known anatomical structure, they present with many different types of pain characteristic, they are sometimes accompanied by quite bizarre features, and they may occasionally be accompanied by unusual personality characteristics. Idiopathic pains also have a tendency to respond poorly to treatment, or even respond adversely to therapeutic measures. These features certainly do nothing to dispel the belief of many that they have a psychogenic origin.

Arner and Myerson (1988) included a group of so-called 'idiopathic pain' patients in a study of responsiveness to opioids. They described these as patients in whom chronic pain could not be accounted for by any demonstrable organic pathology (Almay 1987; Williams and Spitzer 1982). There is a disproportion between suffering and physical signs, even if the pain originally started following tissue damage. They suggest that although these patients are not suffering from recognized psychiatric disease, there are abnormal personality traits and the development of abnormal pain behaviour. Such patients have often undergone extensive investigation and had multiple medical consultations. This may lead to a distortion of symptoms, and frustration, cancerophobia, and secondary neurotic symptoms may develop.

In the Arner and Myerson (1988) study, it was found that the intensity of this type of pain did not change significantly when the patients were given opioid analgesics, compared with placebo. This was in contrast to patients suffering from known nociceptive pain. They were also able to demonstrate a relative opioid insensitivity of neurogenic pains.

There is always the risk that a diagnosis of idiopathic pain may obscure a serious underlying pathology. The typical example of this is the patient who has for a long time complained of perineal pain. As the symptoms may be expressed in unusual terms, and perhaps modern imaging techniques have failed to reveal pelvic pathology, the pain is described as idiopathic, and an underlying malignant condition only becomes apparent when extensive invasion of the sacral plexus has occurred. Fortunately, this situation is relatively uncommon, but care to exclude organic disease must be meticulous and the situation reviewed from time to time when a diagnosis of idiopathic pain is considered. Conversely, encouraging a belief in undiscovered disease is harmful to the patient with a true idiopathic pain syndrome.

Sicuteri (1981) has proposed the idea of non-organic central pain to account for some common syndromes, such as migraine, and a more generalized whole body pain syndrome which he terms 'central panalgesia'. There seems to be some central mechanism which is active before vasodilatation or constriction in the evolution of a migrainous headache. He suggests the presence of a neurotransmitter dysfunction, possibly involving 5-HT, resulting in disorder of the antinociceptive system. He proposes that there are some similarities between migraine and the abstinence syndrome, and the role of clonidine in suppressing both tends to support this connection. Is there a chronic deficiency of endorphins? Samples of CSF taken during migraine attacks have been shown to exhibit low levels of endogenous opioids. Perhaps there is a 5-HT supersensitivity, or an opioid receptor subsensitivity.

In Sicuteri's syndrome of panalgesia, there are more systemic effects: pain in the limbs and trunk as well as the head, 'psychic irritability', sensory hyperaesthesia, and vegetative disturbances (nausea, vertigo, constipation, flushes, and so on). There is an elevated sensitivity to 5-HT, dopamine, and tyramine and on injection of parachlorophenylalanine (a 5-HT synthesis inhibitor), there is provocation of pain in subjects who are headache sufferers, but not in other individuals.

Sicuteri's proposals are controversial, and require much more evidence for them to be fully accepted. In the future these ideas could become part of the explanation for some of the currently more inexplicable pain syndromes in which there is no known pathology, including idiopathic pains in the head and face, as well as the whole limb and whole body pains which are frequently ascribed to psychopathological causes, and yet in whom no recognizable psychopathology can be demonstrated. However, many of the symptoms described in 'central panalgesia' are the symptoms common to various psychological disorders.

Orofacial pains

Orofacial pains have been variously classified by different writers depending upon their own particular field of interest. Mumford (1980) divides atypical facial pain into three separate categories: those that have poorly understood pathological diagnosis, those that result from augmentation of normal stimuli, and those resulting from psychological illness.

Mumford's principal differential diagnoses in the category of incompletely understood conditions include the following:

(1) Toothache, due to dentine defect or pulpitis—a number of minor dental defects may lead to summation of subthreshold neural impulses, resulting in clinical pain;

(2) impacted teeth;

(3) lateral periodontal abscess;

(4) sinusitis;

(5) acromegaly;

(6) Paget's disease;

(7) arthritis of the neck;

(8) muscular pain, mainly in the masticatory muscles. This may result from stress and cause tooth grinding, or bruxism. This pain and tension can spread to the temporal region and neck, developing into a widespread myofascial syndrome;

(9) vascular pain (temporal arteritis, migraine, cluster headache);

(10) neoplasia, especially of the maxillary sinus, the nasopharynx, and the cervical vertebrae;

(11) nerve damage, following facial injuries, tooth extraction, postherpetic neuralgia, anaesthesia dolorosa;

(12) post-incisional pain following surgery, such as Caldwell-Luc procedure;

(13) neuralgias.

It must be borne in mind that psychogenic disorder does not confer immunity from organic disease.

Migraines and cluster headaches can present in atypical ways. However, an attack of migraine rarely lasts more than a few days, and cluster headaches are usually measured in hours. The former are frequently accompanied by nausea and photophobia, and the latter are usually characterized by an ipsilateral red, watery eye and a blocked nose. The time course and accompanying features of these conditions usually differentiate them from the other causes of chronic facial pain.

Bell (1989) attempts to differentiate between chronic facial pain which has a neurogenic origin, and that which has a somatic origin. He cites the following as evidence that a pain has a neurogenic origin:

1. The pain often has a burning quality, but may be spontaneous or triggered, or unremitting.
2. The pain may occur to a degree which is disproportionate to the stimulus.
3. It may be accompanied by other neurological symptoms.
4. It may be initiated or accentuated by efferent sympathetic activity in the area.

Neurogenic pain is then divided into two major categories: paroxysmal neuralgia and deafferentation pain. These are primary and secondary conditions, and only the primary conditions can be blocked by local anaesthesia at the site of the pain.

Of somatic pain, Bell states that musculoskeletal pains relate to bio-mechanical function, and that these demonstrate a graded reaction which is proportional to the applied stimulus. Visceral pains are unrelated to bio-mechanical function and do not generally respond to provocation until a certain threshold has been reached.

As do most writers in the field of orofacial pain, Bell stresses the importance of differentiation between nociception, pain, suffering, and pain behaviour, and emphasizes the emotional significance of the face and mouth in normal and abnormal body image.

The concept of subjects who experience facial pain as an augmentation of normal external stimuli has been explained as the facilitation of transmission of neural impulses by decreased inhibition (Mumford 1980). Many such individuals suffer from 'cancer phobia', and their disease conviction has usually been reinforced by excessive medical investigation. There is little more that can or should be done for these patients other than to reassure them of the benign nature of their symptoms, and perhaps to encourage an increase in facial and oral activity, so as to stimulate large fibre activity.

Having eliminated the above causes for facial pain of unclear aetiology, there remains a large number of patients who fall into the perhaps unsatis-factory category of psychogenic facial pain. These patients, whatever the origins of their pain, undergo real suffering and should have their symptoms taken seriously, even though the management may not involve surgical or 'medical' treatment. In many cases the apparent neuroticism presented may be a result of, rather than a cause of, the pain (Sternback and Timmermans 1975).

Psychogenic facial pain usually has a non-anatomical distribution, and no identifiable cause. The pain is often variable, and may even move to other areas or be accompanied by other strange symptoms. Three-quarters of such patients are female, and a third complain of other sensory symptoms, such as anaesthesia, paraesthesia, or hyperaesthesia. Over half are found to suffer from headaches, and 73 per cent from depression or anxiety (Mock *et al.* 1985). Most of these patients have been extensively investigated, treated, and have usually lost many of their teeth in attempts to alleviate the symptoms. The pain is often described as being deeply seated and initially unilateral, but it may later become bilateral, and is nearly always poorly localized. It is often described by its affective qualities, such as 'crushing, tearing', as well as its constant, aching nature. Having been present for many years, its origin is often ascribed by the patient to some minor injury or operation, or it may relate to some specific life crisis. Attempts to quantify the pain using analogue scales often result in the patient giving it the maximum score on the pain scale, at whatever stage of the condition, or despite other evidence to the contrary. The pain is frequently described as being worse following therapeutic intervention (Soloman and Lipton 1988). Similar descriptions can be applied to atypical odontalgia, glossodynia or burning mouth syndrome, and some of the intractable perineal and genital pains. To many individuals, the genitals and rectum have an equally important role in body image as does the face!

It is often characteristic that such patients have a desire for surgery, and refuse psychiatric help. They usually deny emotional conflicts. It has been said (Soloman and Lipton 1988) that before the onset of pain they had personalities that are usually described as workaholic, but once the pain syndrome has developed, they are prone to dysphoria, insomnia, and fatigue. They lose pleasure in their work, social and sexual lives. They exhibit an inability to cope, and frequently suffer a disintegration of family and social life.

Some would include temporomandibular joint syndrome with the other types of facial pain presumed to be psychogenic. This condition, with facial pain aggravated by chewing, a decreased range of motion of the jaw, and tenderness of the muscles of mastication, possibly has many causes and does in many situations respond to physical measures. However, there seems to be a number of patients in whom this condition follows the same pattern as other facial pains of unknown, idiopathic, or psychogenic origin.

Asymmetry of facial blood flow associated with atypical pain, and changes in facial blood flow in response to pain, have been reported (Drummond

1988) from thermographic studies, but this does not necessarily confirm an organic cause for these pains, and most reviews favour a psychogenic cause.

Bell (1989) differentiates the chronic orofacial pains into those which he calls 'chronic structural pain', and 'chronic functional pain'. In the former, there is evidence of a nociceptive input, but the pain escalates with time even if there is no increase in noxious stimulus. He suggests that there may be a decrease in endorphins with time, accompanying the increase in suffering. However, even in the 'chronic structural pain', chronicity changes the clinical characteristics of the pain syndrome so that non-anatomical and non-physiological behavioural changes occur. As this happens, the different syndromes become more alike and management methods less diverse. He suggests that some of these cases can be helped by a combination of anti-nociceptive therapy and supportive psychotherapy. 'Chronic functional pain' becomes more severe with time, even though peripheral input may drop to a very low level, and local antinociceptive therapy offers no benefit. Local anaesthesia may produce a temporary placebo effect, but the relapse tends to reinforce the symptoms and escalate the suffering. The pain ceases to bear any relationship to the original area of input and pain sites become multiple and bilateral. Response to treatment in this category of pain may be strange and unpredictable.

Although it has often been supposed that chronic facial pain may be related to recognized social and psychological characteristics, investigations by Gerschmann et al. (1987) are unable to confirm this proposition. They do, however, demonstrate that the outcome of treatment in this syndrome is related to the degree of past treatment which the patient has received and suggest that the experiences of patients at the hands of the medical and dental professions may play a role in the maintenance of chronic pain, and that there is a strong behavioural component. They stress that we should aim to minimize iatrogenic damage.

Management

Treatment of these idiopathic pain syndromes is often unrewarding. There is generally a poor response to analgesics, although their consumption is often heavy in such patients. Nerve blocks, acupuncture, and biofeedback have all been reported as beneficial, although the response is often short-lived. Surgery is usually unhelpful, and usually harmful, often aggravating the condition as well as introducing new iatrogenic problems. The only treatments which are frequently quoted as being of use are cognitive coping therapy, and the prescription of tricyclic antidepressants. The former may offer long-term hope to these afflicted patients providing that there is sufficient motivation and a willingness of the patient to accept that psychological techniques may be better than radical invasive procedures.

The tricyclic antidepressants have been found to be the most useful drugs in the management of the idiopathic pains, although the monoamine oxidase inhibitor antidepressants may have a use in this field. Amitriptyline (Sharav *et al.* 1987), doxepin, and dothiepin are the agents of which most experience has been gained. Individuals differ in their susceptibility to the side-effects of each drug, and it may be necessary to try more than one of them before there is acceptance. The dose should be started at a very low level, and increased gradually over a period of weeks. The effective dose varies considerably between individuals, and there may be a therapeutic 'window' above and below which the drug is less effective. The dose should be increased until a therapeutic effect is obtained or unacceptable side-effects supervene. It seems that the analgesic properties of these drugs are not related to their antidepressant effects. They possibly act by raising central serotonin levels with the corresponding enhancement of brainstem and spinal inhibitory mechanisms (Feinmann 1985).

Feinmann, Harris, and Cawley (1984) performed a double blind trial in patients with psychogenic facial pain to compare the use of dothiepin and placebo, both with and without the use of a nocturnal biteguard. Their patients were assessed as having no psychiatric diagnosis (43 per cent), non-depressive neurosis (22 per cent), or depressive neurosis (35 per cent). They found that the biteguard was abandoned by 38 per cent of patients by 3 weeks, and by a further 25 per cent by 9 weeks. This device was not found to be effective in relieving the symptoms. On the other hand, of those taking dothiepin (average dose of 130 mg/day over 9 weeks), 83 per cent were able to significantly reduce their intake of analgesics, compared with only 42 per cent of those taking placebo therapy. It was found that this beneficial effect could be maintained, but required long-term continuation of medication at the maximal dose. It is suggested that psychogenic facial pain arises at sites of vasodilatation within muscles or the capsules of the temperomandibular joints. They agree with the suggestion (Sternbach 1976) that the pain may result from depletion of brain serotonin, especially in the dorsal raphe nucleus. Tricyclic antidepressants potentiate endogenous opioid systems, and hence decrease sensitivity to painful stimuli.

Other pains without obvious pathology

Chronic pain without any demonstrable pathology, and which is atypical of the types of pain seen after damage to the nervous system, may present in any part of the body but, apart from the lack of a proven pathology, these syndromes also have in common that they have all at some time been described as psychogenic. Perhaps they represent an abnormal augmentation or perception of common physiological variables. Syndromes such as irritable bowel syndrome, although usually described as psychogenic, have had a physiological mechanism described to account for the presenting symptoms. Others, such as those in patients complaining of whole body or whole limb pain, are more difficult to explain. Most of these syndromes are best managed by training patients to control stress, particularly by the use of cognitive behavioural training, and sometimes with the help of psychotropic medication, especially the antidepressants.

One relatively common syndrome, often seen by gynaecologists but occasionally referred to the Pain Clinic, is chronic pelvic pain without obvious pathology (CPPWOP). This is described as a female affliction, although possibly some of the male patients complaining of chronic testicular pain may represent a similar condition. These patients complain of pelvic pain, usually deeply felt, but into which extensive investigations have failed to reveal any pathology to account for the symptoms. There is usually a deep conviction of disease in the pelvic organs or genitalia, and a long history of seeking medical or surgical cures. Many have regarded this condition as a form of neurosis, whereas others have put forward a wide range of suggested physiological explanations for the pain. One of the most popular theories has been that of pelvic congestion caused by vasomotor instability, with venous distension and oedema. Another suggested is microscopic endometriosis, or the persistence of abnormal sensitivity in pelvic viscera following previous parametritis. All proposals are lacking in firm evidence.

Many psychological studies have been undertaken of women with CPPWOP (Renaer and Guzinski 1978) and have suggested a correlation between the pain and the onset of stressful situations, particularly where the patient's female identity was under stress.

The differential diagnosis of chronic pelvic pain should include consideration of scarring and adhesions (although the term 'adhesion is often used as a convenient label for an unknown pathology), a distortion of normal pelvic anatomy, loss of pelvic support, endometriosis, and vascular abnormalities. Consideration should also be given to disorders of associated systems namely the urinary, gastrointestinal, vascular, nervous, and musculoskeletal systems. Psychological mechanisms are undoubtedly an important factor in many patients, and psychological help may be a useful approach in a condition where surgery often provides only temporary relief, and other medical interventions have a disappointing outcome.

References

Almay, B. G. L. (1987). Clinical characteristics of patients with idiopathic pain syndromes. Depressive symptomatoiogy and patient pain drawings. *Pain*, **29**, 335–46.

Arner, S. and Meyerson, B. A. (1988). Lack of analgesic effect of opioids on neuropathic and idiopathic forms of pain. *Pain*, **33**, 11–23.

Bell, W. E. (1989). *Orofacial pains* (4th edn). Year Book Medical Publishers Inc., Chicago.

Drummond, P. D. (1988). Vascular changes in atypical facial pain. *Headache*, **28**, 121–3.

Fienmann, C. (1985). Pain relief by antidepressants: possible modes of action. *Pain*, **23**, 1–8.

Fienmann, C., Harris, M., and Cawley, R. (1984). Psychogenic facial pain: presentation and treatment. *British Medical Journal*, **288**, 436–8.

Gerschman, J. A., Hall, W., Reade, P. C., Burrows, G. D., Wright, J. L., and Holwill, B. J. (1987). The determinants of chronic orofacial pain. *The Clinical Journal of Pain*, **3**, 45–53.

Mock, D., Frydman, W., and Gorden, A. S. (1985). Atypical facial pain: a retrospective study. *Oral Surgery, Oral Medicine, Oral Pathology*, **59**, 472–4.

Mumford, J. M. (1980). Atypical facial pain. In *Persistent pain: modern methods of treatment*, Vol. 2 (ed. S. Lipton), pp. 249–82. Academic Press, London.

Renaer, M. and Guzinski, G. M. (1978). Pain in gynecologic practice. *Pain*, **5**, 305–31.

Sharav, Y., Singer, E., Schmidt, E., Dionne, R. A., and Dubner, R. (1987). The analgesic effect of amitriptyline on chronic facial pain. *Pain*, **31**, 199–209.

Sicuteri, F. (1981). Persistent non-organic central pain: headache and central panalgesia. In *Persistent pain: modern methods of treatment*, Vol. 3 (ed. S. Lipton and J. Miles), pp. 119–40. Academic Press, London.

Soloman, S. and Lipton, R. B. (1988). Atypical facial pain: a review. *Seminars in Neurology*, **8**(4), 332–8.

Sternbach, R. (1976). The need for an animal model for chronic pain. *Pain*, **2**, 2–4.

Sternbach, R. A. and Timmermans, G. (1975). Personality changes associated with reduction of pain. *Pain*, **1**, 177–81.

Williams, J. B. W. and Spitzer, R. L. (1982). Idiopathic pain disorder. A critique of pain-prone disorder and a proposal for a revision of the DSM-1 category psychogenic pain disorder. *Journal of Nervous and Mental Disease*, **170**, 415–19.

10 Psychological management of pain

Pain is a subjective experience. It therefore follows that organic pathology cannot be the sole factor in determining how that experience is perceived by the sufferer. Important psychological and environmental factors will interact with any physical factors to alter that experience, and the way in which this interaction occurs will vary between individuals. There will also be wide variation in the behaviour resulting from the experience of pain. Personality traits of the sufferer, and the reactions of others to an individual's suffering will have an influence on the pain experience, as well as the significance which that pain seems to have to the individual.

Any chronic illness or disability including pain, will affect the emotions and behaviour of the individual. Pain often leads to anxiety and tension, which themselves may increase the experience of pain. The resultant loss of ability to cope with the situation, leads to loss of self confidence, increased avoidance of others, and withdrawal, increasing the helpless and hopeless plight of the sufferer. Feelings of depression frequently result, perhaps accompanied by feelings of guilt, loss of self esteem, self pity and even anger.

Neurotic disorders such as anxiety, depression, or exaggerated preoccupation with bodily function, are often associated with pain syndromes. Whilst it is accepted that conditions such as tension headache, some types of back pain and irritable bowel syndrome may be a result of these processes, the emotional disturbance of a pain sufferer may actually be a result of prolonged suffering. Certain psychoses may manifest as a pain problem, most notably severe endogenous depression, with hypochondriacal delusions. There is certainly a strong likelihood that some chronic pain conditions are associated with disturbances of neurotransmitters, in a similar way to that proposed for some psychoses. However, psychotic disorder is not frequently seen in the average pain clinic, and the decision of many doctors to refer to a psychiatrist a patient with a pain which has defied all diagnostic attempts, is often counterproductive. In most of these cases the psychiatrist will report that no evidence of psychiatric disorder in that patient was found, and the belief that the patient's mood disturbance will improve when pain is controlled. Unfortunately, many patients interpret this referral as an indication that the doctor disbelieves the patient's complaints of pain, and because no cause for it has been found, the patient must, by a process of exclusion, be mad. Most patients can see no connection between emotion and pain or other bodily sensations, and therefore assume that the doctor supposes that the pain is imagined. This may also suggest to the patient that all further attempts to investigate or relieve the pain have been abandoned.

The situation is further complicated by the recognition that some patients with chronic pain exhibit 'abnormal pain behaviour'. It is difficult to know whether this is a personality trait of that individual which finds expression in the manipulative behaviour exhibited in other aspects of their lives, or whether the behaviour results from an unconscious attempt to express suffering and to achieve secondary gain. The expression of suffering, disability, and the adoption of the 'sick role' can be used to obtain further unwarranted medical attention, to achieve emotional control over other members of the family and thus gain love and attention that may be otherwise denied, to avoid or excuse unpleasant aspects of the patient's life (such as inability to cope with employment, domestic stress, or emotional and sexual relationships), or to vent the anger and frustration of continued disability.

The evaluation of the relationships between pain, personality, psychotic and neurotic disorders, and pain behaviour is often extremely difficult. The differentiation between pain and suffering is an important step in determining the most appropriate management of the patient with chronic pain, since although the two may be interdependent, the treatment of each may need to proceed along different lines.

The problems outlined in the preceding paragraphs may exceed the skills and experience of many Pain Clinic doctors, and the involvement of a clinical psychologist with a particular interest in pain, is a major asset to an effective Pain Clinic. When setting up a Pain Clinic, this should be stressed to those responsible for providing resources; for a busy Pain Clinic, two sessions of a clinical psychologist's time dedicated to treating pain patients is probably the minimum necessary. In the long term this may be more cost-effective than many physical and pharmacological therapies, and may represent the only help that can be offered to many patients for whom medical therapy has been unsuccessful. An effective psychological service may help many dependent patients to return to a more independent and productive lifestyle.

Unfortunately, clinical psychology services are not always available to a Pain Clinic. Although this chapter is not intended to be a 'do-it-yourself' guide to being a clinical psychologist, it aims to point out some of the ways in which psychologists have contributed to the management of chronic pain, and perhaps enable the Pain Clinic doctor to see where some of the techniques developed by psychologists can be used in everyday pain management.

The role of a clinical psychologist in the Pain Clinic can be summarized as follows:

(1) counselling and support of the patient to enable the patient to communicate more effectively with the medical profession and other

professionals, as well as coping with the interpersonal stresses and problems arising both at home and socially;

(2) education of the doctor in the psychological needs and problems of pain patients;

(3) assessment of psychological parameters, outcome of treatments, and social/environmental factors with a bearing on the patient's pain;

(4) investigation and research into chronic pain;

(5) treatment.

Many individuals suffering from chronic pain have lost the ability to cope effectively with their disability and distress, as well as the ability to communicate effectively with professional medical and paramedical staff. This may have been compounded over the years by frustrated doctors who have dismissed the patient's complaints as imaginary or incurable. The patient is sometimes made to feel to blame for their own predicament, thus increasing feelings of guilt and inadequacy. The family stresses that result from chronic pain and disability, with avoidance of normal family, social, and sexual activity, are often difficult for the patient to understand. The relationships between mood, emotions, personality, and pain are rarely appreciated by patients. Although an undisguised psychoanalysis of the patient's personality, motives, and behaviour may not be appreciated by many patients, and in fact may be interpreted by many as a statement on their sanity, the clinical psychologist can often help a patient to gradually appreciate how emotions affect bodily sensation and function and that this is not an accusation. Simple analogies can be used to illustrate this relationship, as applied to other areas of the patient's experience.

A psychologist can often help the patient to reach a better understanding of the nature of chronic pain. Knowledge of how pain can be influenced by mood and how behaviour can be altered by pain, sometimes inappropriately, may help the patient to come to terms with problems in a more realistic way, and by modifying abnormal pain behaviour, improve relationships with other individuals in the family or outside. Patients should also be helped to understand the negative effects of seeking the ultimate and unlikely 'cure'. Further counselling from an appropriately-trained psychologist may also enable the helpless patient to regain some assertiveness in a positive direction, with therapy aimed at improvement of self image and developing positive aims in life.

Educating the doctor is also important; appreciation of psychological factors does not always come naturally to members of the medical profession. If resources are available, there may be some benefit from a psychologist being present at the initial interview with the patient in a Pain Clinic. A discussion after the interview may enable the psychologist to point out some of the psychological factors influencing the patient's pain perception and abnormal pain behaviours, or help to identify patients who should be assessed by a psychiatrist. If the psychologist is seen to be a member of the Pain Clinic staff, rather than as a 'sort of psychiatrist' who is usually based in a psychiatric hospital, the patient is less likely to erect barriers and to reject psychological assessment and treatment for what they perceive as a purely physical condition.

Psychologists have provided important means of assessing patients in pain. As mentioned above, the psychologist can help to select those patients who may have a psychiatric disorder which is a major factor in the cause of the symptoms, or where the symptoms and behaviour are thought to be disproportionate to the objective signs of disorder. Many of the formal questionnaires designed to assess personality, mood, illness behaviour, and coping ability have been designed by psychologists and although they are often suitable for scoring by non-psychological staff, their full interpretation and significance can be best appreciated by a professional psychologist.

The measurement of pain is a difficult, if not impossible, task but there are various techniques available which help in the assessment of the effects of pain. These have been dealt with in a little more detail in Chapter 3. The questionnaires developed by psychologists can produce scores for illness behaviour, the impact of disability, mood, and patient attitudes to the responsibility for control of their pain, which can provide valuable information in assessing the effects of treatment in the Pain Clinic. In this way a clinical psychologist can monitor a patient's progress and evaluate new forms of treatment.

The problem of providing appropriate medical intervention in the management of pain requires an assessment of illness behaviour, and whether the patient is exhibiting appropriate symptoms. Waddell *et al.* (1984) were able to demonstrate that the amount of treatment that patients with back pain received bore more relationship to the distress and illness behaviour that they exhibited than to physical disease, and that the successful outcome of physical treatment was related only to the latter. They listed a series of inappropriate symptoms and signs in back pain which signal poor prognosis from physical intervention. Inappropriate symptoms are: tailbone or whole leg pain, whole leg numbness, whole leg giving way, no intervals free of pain, intolerance of treatment, and a series of emergency admissions related to episodes of pain. Inappropriate signs are: superficial, widespread, and non-anatomical distribution of tenderness, pain on stimulated rotation and axial loading, straight-leg-raising pain which improves with distraction, regional distribution of sensory and motor deficit, and over-reaction to painful stimuli (crying out, facial expression, muscle tension, sweating, or collapsing).

Behavioural approaches to chronic pain

Pain behaviours communicate that a person is suffering, and these may include: speech, facial expression, posture, requesting and taking medication, seeking health care intervention, and refusing to work. The application of operant conditioning methods of pain management was examined by Fordyce (1973), who distinguished between respondent pain behaviours, which occur in response to a nociceptive antecedent stimulus, and operant pain behaviours, which are sensitive to consequences and available to the conditioning effects of contingent reinforcement. Such conditioning effects would be perhaps the pleasant effects of analgesics, attention and concern from family and friends, or the avoidance of unpleasant responsibilities. In some environments, pain behaviours may be conditioned to exist, and be responsive to social control, long after healing has occurred and for reasons different from those acting at the time of original injury. Hence the patient may find that illness behaviour is rewarded by increased attention and sympathy. The patient with chronic pain learns to avoid tasks which have previously been associated with an increase in pain, although this expectation may no longer be justified. Also, the patient in pain rarely has well behaviour rewarded, as this leads to disbelief by others of the reported level of suffering. Illness behaviour is rewarded by being told to 'take it easy'. The aim of operant conditioning is to identify sickness behaviours and remove the gains, whilst also identifying well behaviour and rewarding this. Operant conditioning can increase a patient's level of physical activity and reduce dependence on medication. It appears that these changes can be maintained in the long term (Turner and Chapman 1982b).

Fordyce (1973) and his associates have developed treatment packages based on these principles. Because of the intensive nature of such programmes, and the need to remove the patient from environmental factors which constantly reinforce pain behaviour, operant conditioning programmes are most frequently organized on a residential basis. This has obvious drawbacks as far as resources are concerned and such a programme generally requires skilled psychological input.

The goals which are set in a behavioural programme can be summarized as follows:

1. Activity levels are gradually increased by reinforcing with social praise, and rewarding with the chance to rest when goals have been met.

2. Medications are gradually decreased, by replacing previous medication with a 'pain cocktail'. This mixture has its analgesic or sedative components gradually reduced, and the cocktail is given on a timed basis, rather than in response to reports of pain.

3. Staff and families are trained not to reinforce sickness behaviour by expressions of sympathy or reduction of work responsibilities, but to positively reinforce and reward well behaviour.

4. The subjective experience of pain is only dealt with by showing disattention to pain behaviour.

Such conditioning programmes are reasonably successful in increasing activity levels in patients with chronic pain and in reducing medication demand, although the reduction of reported pain levels is often minimal. Patients should be told to expect an improved quality of life rather than an absolute reduction in their pain.

Cognitive therapy

Cognitions (attitudes, beliefs, and expectations) related to certain situations may influence emotional and behavioural reactions to those situations. Cognitive therapy is aimed at altering the experience of pain by modifying cognitive variables. This may involve distraction techniques, imagery, or altering the significance of the pain to an individual. Unlike operant conditioning, cognitive therapy does not ignore the experience of pain but aims to reduce the suffering that it produces in the patient by altering the way in which pain signals are interpreted by that individual.

Under the broad terms of 'cognitive therapy' patients are taught to identify negative beliefs concerning pain and its effects, and to substitute these with positive thoughts and actions. Imagery, distraction, and relaxation techniques are taught as coping strategies for dealing with the stresses associated with chronic pain. If the patient can be taught to be more aware of events which increase or decrease pain, he can develop a greater feeling of control over the pain instead of continuing with a helpless and hopeless attitude to symptoms. An important aspect of cognitive therapy is the use of relabelling techniques, whereby the patient is taught to reinterpret the sensations of pain. This is discussed later.

Although intensive behavioural or cognitive therapy is heavily dependent on the clinical psychologist's time available, the basic ideas are frequently modified and combined with other techniques as part of a pain management programme.

Fernandez (1986) refers to a trimodal system of pain management: cognitive strategies, behavioural manipulations, and physical intervention. The first of these is a system where the point of input is the patient's mind, and the patient can learn to manipulate this input for himself and therefore develop a greater sense of independent control. Behavioural manipulations, such as operant conditioning, biofeedback, and hypnosis require input by the therapist, although the patient can learn to have some control over this input. Physical interventions are seen as the method with the least autonomy for the patient, and those requiring maximum external control. The perception by the patient of the locus of control is seen as an important factor in the long term successful management of chronic pain.

Patients can learn to use imagery in a variety of ways to control pain. The attention can be focused on an image which is emotionally incompatible with pain, such as a pleasant event in the patient's personal or professional life, perhaps more readily conceived, on images which are sensorily incompatible with pain. An example of the latter would be images of warmth, sunshine, soft warm sand, and so on for the patient who finds warmth and relaxation a natural ease for their symptoms. Alternatively, patients can be encouraged to transform the pain to an alternativea and less threatening image, such as the acceptable discomfort following vigorous exercise, or some other exciting activity.

An important addition to the imagery techniques is the use of the patient's self statements, which he uses to enable him to cope with the stressful situation. This does not aim to reduce the painful experience, but to improve the patient's ability to cope with it. Statements must be positive and self reinforcing, concentrating, for example, on ideas such as 'You are in control', or 'Relax, you can cope with this problem'. Obviously, such phrases have to be suitable for the individual patient.

A further method which is often advocated, is the attention–diversion method. Patients are taught to redirect attention away from the painful experience. This may be simply to visions of other more pleasant activities, but may involve active attention-demanding activities such as problem-solving or mathematical exercises. We have found most patients to be particularly sceptical about this latter approach and it is important not to suggest banal activities which antagonize the already frustrated patient. An example of an inappropriate attention-diverting activity which was suggested to a group of pain sufferers was to concentrate on counting the holes in ceiling tiles. This suggestion met with the derision it deserved. However, when it is pointed out to patients (with relevant examples) how difficult it is to concentrate on two thought processes at the same time, many begin to see that stimulating diversion can be useful. Many patients are able to carry this theme further by becoming absorbed in hobbies which they enjoy. This can be combined with realistic goal setting exercises, and consideration of activities which can be physically accomplished with some effort, compared to perhaps previously enjoyed activities which cannot be attempted without frustration and further pain.

Relaxation and biofeedback

There is a common assumption that many types of chronic pain are associated with, and may have a causative factor in, abnormal levels of muscle tension. This is certainly a convenient explanation to the patient for symptoms which are otherwise difficult to understand. However, there is little evidence that muscle tension is the cause of chronic pain syndromes. Possible exceptions are certain types of headache and some tempero-mandibular pains. Nevertheless, chronic pain syndromes are often accompanied by tension, either physical or emotional, and this may aggravate the chronic pain, reducing receptiveness to other coping strategies. A patient who has learnt a technique for deep relaxation is generally more able to cope with chronic pain, and may find that the technique can even lead to a slight reduction in pain when it is most severe. Relaxation training therefore commonly forms a part of many other progammes of psychological pain management (Turner and Chapman 1982a).

Biofeedback utilizes electronic monitoring of physiological functions to enable the patient to recognize what are supposedly undesirable features, and to learn to exercise some conscious control over what are normally unconscious functions. Examples are: monitoring the EEG to enable the subject to increase alpha brain activity, cephalic blood volume pulse feedback, and EMG feedback to reduce muscle tension. Although there has been success using biofeedback techniques in managing tension headache and temperomandibular pain, evidence of success in other chronic pain syndromes is scanty. Consideration should be given to commercial interests resulting from the elaborate equipment required for biofeedback when evaluating some of the claims made for this form of treatment.

Relaxation training takes two main forms. Perhaps the most widespread method involves the patient in conscious tensing and relaxing of all muscle groups in a systematic and sequential manner. This takes a considerable amount of time and may be difficult or painful for many patients. Of greater value and ease for most patients, is mentally-induced relaxation, which follows when the patient concentrates on relaxing imagery, control of breathing, and more generalized concepts of tension flowing out of the body.

Hypnosis

The skills of hypnotherapy need to be learnt to a large extent by practical example, although there are many excellent books which give an insight into, and practical advice on, the use of hypnotherapy in pain management (Elton et al. 1983; Scott 1974; Hilgard and Hilgard 1975). Hypnosis undoubtedly has a strong placebo effect in some individuals but, certainly in highly hypnotizable subjects, this is not equivalent to the hypnoanalgesia that can be developed. The perception of pain can be altered by hypnosis, as is adequately demonstrated with hypnoanaesthesia during surgical procedures, although the physiological changes associated with that pain may not be suppressed. Even in subjects who are not sufficiently susceptible to hypnotic suggestion, the technique is useful to alter the effective component of the pain experience, and for enhancing imagery and relaxation techniques. Hypnotherapy is rarely a means of completely controlling symptoms of chronic pain in the long term, but is commonly a useful adjunct to other cognitive and behavioural manipulations. Hypnotherapy should not be viewed as a means of introducing complex psychotherapeutic measures in the management of chronic pain by the 'enthusiastic amateur'.

Hypnosis has a valid and widespread application in the treatment of acute pain, in such situations as dentistry, or burns dressings, but in chronic pain syndromes there are many factors influencing the perception of pain and pain behaviour. Sometimes when the chronic pain has an obvious organic cause, as in malignant disease, hypnosis can be directed at pain reduction, although consideration must be extended to other psychological factors which have a bearing on that patient's experience. Hypnosis may here form a valuable part of the overall aims in helping the patient to cope with the symptoms. However, in chronic pain where there is little or no evidence of physical abnormality, then hypnosis is less likely to be effective in providing analgesia as a simple aim (Orne and Dinges 1984).

Hilgard and Hilgard (1975) suggest three main approaches to the use of hypnosis in pain management: direct suggestion of pain reduction, alteration of the experience of pain, and diversion of attention from the pain. A similarity is immediately seen between the last two of these approaches, and the aims of cognitive therapy as described earlier, thus illustrating how hypnotic techniques can be used to facilitate other psychological strategies.

Various imaginative techniques have been used by different therapists to reduce the experience of pain under hypnosis. The only limits are the respective imaginations of the therapist and the patient. However, relatively simple concepts can be employed, such as the image of warmth or cold in the painful area, visions of pain-reducing medication flowing through the painful area, gradual diminution in size of the painful area so that it reaches small and manageable proportions, or the image of a 'pain meter' in which the patient causes the needle to gradually move down from the 'penalty' zone to the 'parking allowed' zone.

Pain management programmes

The psychological approaches to chronic pain management, as outlined above, are often utilized as part of a pain management programme. This enables a combination of techniques to be employed in a structured programme for a small group of patients. Again this type of treatment is very demanding of personal contact time, but by organizing group therapy, the time of the psychological, medical, and physiotherapy or other professional staff can be used optimally.

The prime aim of a pain mangement group is not pain reduction, although this sometimes results as an added bonus. The main objectives are: to increase self perceived control over pain, to increase self perceived independence, to increase levels of physical and social activity, and to reduce levels of emotional distress.

It is important that the patient's understanding of his or her condition is improved. A patient who constantly refers his or her chronic pain symptoms to the classical medical model is doomed to frustration and disappointment with the results of medical treatment, which may be manifest as anger against the medical profession. These reactions are counterproductive in coping with pain and the stresses that result from it. Some knowledge of the nature of chronic pain, its differing nature from acute pain, and the possible reasons why it has not responded to medical treatment, is a useful beginning to any pain management programme. It may be advantageous for this part of the programme to be undertaken by the Pain Clinic doctor, so that patients feel that this new approach to their pain has continuity with previous medical contacts, and to validate the psychological approach in the eyes of some of the more mistrustful patients.

There is a focus on behaviour associated with pain. Patients are encouraged to examine their physical abilities, and by a series of graded exercise programmes, to increase their overall levels of activity. By setting realistic goals, and then attaining these goals, patients can gain a tremendous increase in self respect, motivation, and independence. The involvement of a physiotherapist and occupational therapist can be invaluable in helping patients to attain goals, and reduce unnecessary handicap.

Psychological adjustment is the other prime area of concentration in a pain management programme. Many pain patients have a sense of passivity and helplessness, expecting another person (the doctor or other professional) to provide pain relief. As long as that other person can relieve the pain, then all other problems and disabilities experienced by the patient will be resolved, or so the patient frequently believes. The aim of therapy is to shift this locus of control. Patients are encouraged to take responsibility for managing their own symptoms, and thus regain independence and self esteem. The patient must be in control of his or her own life, rather than the pain being the controlling factor. A major part of this aspect of pain management is to train the patient in pain coping strategies which they can use for themselves at any time, thus giving them independence of the medical doctor–patient relationship. The patient has to be taught that the pain is a part of his or her life, and that it is up to him or her to make that part as acceptable as possible.

Patients are selected by the Pain Clinic medical and psychological staff as potentially suitable for a pain management group. Suitability should not depend on the fact that all other treatment has been unsuccessful as its major criterion. Patients must be motivated with a desire to achieve a degree of independence in the control of their lives; those who are suspected of being excessively passive, dependent, or having a sense of entitlement will not generally do well in a pain management group. On the contrary, such patients may sabotage the benefits for the rest of their group, using the meetings as an opportunity to dwell upon their misfortunes, evoke the sympathy of other members of the group, and as a 'back door' to obtaining even more medical attention. Patients with limited intellect or command of English would also be unsuitable, being unable to appreciate the basic concepts of the group. This may preclude some elderly or very physically infirm patients who similarly would not be capable of the demands expected of them by this type of management. Major personality or psychiatric disorders are also considered unsuitable attributes of patients selected to attend a pain management group. Ideally, having discussed the nature of the programme with the doctor, the patients being considered for this type of management should be individually assessed by the clinical psychologist.

The size of a pain management group is important. It needs to be large enough to enable individuals not to feel excessively exposed and for free-flowing discussions to take place, but not so large that patients feel intimidated or ignored. The ideal number is perhaps eight to ten people. Staff will consist of the clinical psychologist at all meetings, who will act as the coordinator, with the addition when required of medical, physiotherapy, and occupational therapy staff. All participants should be introduced at the initial session, as it is important that patients learn quickly to relax in this potentially threatening situation. Also at the first meeting, the general aims of the course can be discussed and 'rules' are set for the course, such as essential regular attendance and a ban on individual medical consultations in the meetings. It may be appropriate at this stage for the medical member to talk about the nature of chronic pain, and the reason why it may not be amenable to medical treatment, as a better understanding of the problems faced by patients is essential if they are to gain from the course. The rest of the first meeting can be used to introduce patients to the concept of relaxation exercises, their role in managing pain, and a first simple exercise to practise.

The second meeting can start with a brief discussion on the first meeting, reiterating the aims of the course. At each meeting a different aspect of pain management and education can be tackled, combined with a graded exercise programme to increase performance and self confidence, and finally a relaxation exercise. Patients must be encouraged to practise what they have learnt when they go home, and to take 'homework' seriously. As so many new concepts are being introduced, some may find it helpful to receive a hand-out to take home covering the main ideas which have been explored each week.

Suitable topics for the weekly discussion include the following:

1. The relationship between pain, disability, and handicap and the way in which these can be dissociated, or at least minimized.

2. Examination of affective as well as the sensory components of pain, and the pain–tension cycle with its consequence of withdrawal, increased pain perception, and disablement. Introducing the concept of reduction of stress, and therefore pain perception, by relaxation and imagery exercises.

3. Explanation of exercise programme, especially with reference to the effects of posture on pain and disability. The importance of graded exercise, starting at a level of less than that of which the patient is known to be capable and then increasing in graded steps. Patients can also be taught the correct techniques for lifting, lying, posture, and so on to minimize pain.

4. Discussions on goals and limits. The problems of pacing (i.e. the refusal to accept limits) on the one hand, and on the other, refusing to try (i.e. test limits). The concept of increasing activity levels, individual decisions about the personal cost and effort of increased pain and the need for new interests and achievement in the light of restrictions imposed by physical disability. Overcoming loss of occupational and family status (and self esteem) with the establishment of new roles and responsibilities, are all important themes to be covered.

5. A discussion on relationships with the medical profession, how to be effective in consultations, and how to be assertive without antagonizing those from whom you seek help, is useful, as many of these patients by their frustration and anger have become ineffective in obtaining what help may be available.

6. An exploration of cognitive coping strategies. Examples will be given later.

These topics are amongst those that have been found valuable in pain management courses, but individual programmes have to be devised to suit the participants, both patient and therapist.

Apart from exercise training and the discussions, an essential component of every meeting is the relaxation exercise. This term, easily understood by patients, is the vehicle whereby not only effective relaxation and stress control is taught, but cognitive coping techniques are developed with the use of imagery, and perhaps simple autohypnosis can be introduced to patients as a means of using these techniques at home.

An explanation of the function of relaxation training can be given to patients along the following lines:

1. Muscle tension can heighten painful sensation. Relaxation is incompatible with tension, and therefore relaxation reduces the amount of pain caused by tense muscles.

2. A great deal of concentration is required to relax when you are in pain, and although you can learn some special techniques to relax under these circumstances, they do take a lot of effort. While you are concentrating on relaxing, you cannot pay so much attention to your pain. Your brain is incapable of concentrating on two things at the same time. Hence, the more of your concentration that is taken up with relaxing, the less your brain can handle information from other sources. Your pain therefore tends to get pushed to the edge of your awareness. You may still be aware of your pain, but because your attention is so fully occupied with relaxation, your pain is not so overwhelmingly important.

3. When you are relaxed, you are not so likely to become anxious and depressed as when you are tense. Relaxation reduces feelings of anxiety, frustration, and tension, and therefore allows you to cope more effectively with your pain.

4. Relaxation exercises may help you to become more aware of muscles that may become tense, and contribute to your pain. Even when you think you are relaxed some muscles may be tense, and becoming aware of this will enable you to recognize when you need to start your relaxation exercises.

5. Relaxation helps with sleep. Loss of sleep is a common problem in chronic pain, and this reduces the ability to cope with pain.

Relaxation can be taught in two ways. The most widely used method is by progressive muscle relaxation. The group members are instructed to lie on comfortable mattresses, or sit in comfortable but supportive chairs for the exercise; physical disabilities may dictate which is the most suitable. Shoes and restrictive clothing should be removed or loosened, and bladders made comfortable before starting! The eyes should be closed, and the instructor should speak in a quiet but firm way so that instructions are clear and definite. Patients are instructed to control their breathing and concentrate on relaxing during every exhalation. Each muscle group is considered in turn, with the patient concentrating on that area of his or her body until there is a conscious awareness that it is becoming relaxed. Special attention is given to areas, such as the jaw muscles, of which the patient may not normally be aware and yet are common areas of tension. Often the patient is instructed to tense each muscle group in turn before relaxing it. This heightens awareness of that muscle group, and enables the patient to become more aware of the difference between the relaxed and the tense state.

The limb being relaxed should feel heavy, so that if someone were to try and lift it up they would have to lift the whole weight, and the limb would flop back when let go. As each muscle group is relaxed, it is accompanied by the mental repetition of the instruction 'relax', synchronized with slow, controlled respirations. This whole routine may initially take up to half an hour or so, but with regular practice, patients will gradually learn to relax more quickly and completely, and the process can be shortened. After a few weeks, patients may be able to take a deep breath, and by saying to themselves 'relax' during exhalation, quickly allow their muscles to relax effectively. If pain is experienced during the tensing stage of the exercise, then tensing of that muscle group can be done gently, but with explanation that pre-existing tension in that area may be contributing to pain, further demonstrating the need for relaxation.

Patients should practise the technique at home, and in a variety of situations, so that eventually they learn to release tension in different daily activities, and not just whilst lying down in the hospital setting. Sometimes groups are provided with a tape recording of the routine to practise at home, although some feel that this delays the ability to learn the technique independently of the therapist.

The other main method of inducing relaxation is through the use of imagery. Again a comfortable position should be attained, and the therapist's voice should be clear, but gentle and relaxed. A brief period can be spent in concentrating on physical relaxation, with special reference to controlled breathing, and relaxation on exhalation. The patient is then encouraged to imagine a pleasant scene. Initially this has to be a scene of the therapist's choosing, care being taken to avoid any scenes known to be unpleasant to any individuals in the group. The scene, such as a tropical beach, a beautiful garden, or a comfortable and luxurious room with a crackling log fire, is gradually described in terms to which the patient can relate on a personal level. Some aspects of the scene can be described in vivid detail to encourage the patient's imagination to develop the scene. Some details are left deliberately vague so that by encouraging the patient to fill in the gaps his own mental image develops a personal reality. With practice, the patient can develop such a vivid picture that the pile of the rug or the sound of the birds can be felt or heard. Positive aspects of the scene can be used to enhance relaxation via a mental, rather than a physical, process. The feel of soft warm sand on the back muscles, the warm glow of the fire, the relaxing sound of a waterfall, and so on are examples of the types of image that can be encouraged as an adjunct to relaxation. It is important that the patient be given clear mental images with which he can identify and then develop the physical sensations associated with that image. The patient can be led from one image to another in succession, each concentrating on a different relaxing physical sensation.

The sense of developing the relaxed state can be enhanced by utilizing hypnotic techniques as an addition to the imagery. For example, the use of 'counting down' to relaxation can be suggested. As the patient imagines himself descending a staircase to the next scene, each step can be counted and be made to signify an additional sense of deepening relaxation. During a relaxing scene, breathing can be controlled, with each breath 'allowing' a deeper stage of relaxation to be achieved. Some patients may actually enter a light trance state, and by their reactions and movements, it can be seen that they are very involved in the image created. It is important that at the end of the session, all patients are told that they will feel quite wide awake when they open their eyes, although the sense of relaxation will continue in a way compatible with resuming normal activities.

When patients are familiar with these techniques, they can be encouraged to practise at home. They will eventually find that by closing their eyes when relaxing at home and imagining a pleasant relaxing scene, they can readily enter the relaxed state learnt at the group sessions.

Relabelling

This is an important coping strategy for patients with chronic pain to develop. The aim is to train the imagination to explore ways of making the pain experience less unpleasant. Patients can be encouraged to change the words that they use themselves to describe the pain. Words such as 'stabbing' and 'scalding' can be replaced with less emotive words such as 'pricking' and 'hot'. Patients may also develop the ability to reinterpret pain sensations, as a result of relabelling. By imagining the pain as a less distressing sensation, such as tingling, warmth, or even numbness, the pain becomes more tolerable. Another activity is to imagine the pain as some physical entity such as a vegetable with spreading roots, or an octopus with tentacles. Once the pain has been given an imaginary physical entity, it may be possible to mentally manipulate the pain. For example, the pain may be made to retract by an imaginary injection, or even to become numb following an imagined injection of highly potent anaesthetic. Sometimes it is possible to make the pain move to another, less threatening part of the body, so that, for instance, a pain in the neck can be made to shrink and migrate to the tip of the little finger, where it is much less troublesome. Such techniques require a lot of practice and motivation by patients, but some can manage to use them to reduce the distress caused by pain. Relabelling methods are most likely to succeed in patients who have become proficient in self relaxation techniques.

Developing mental skills for coping with pain

Patients can be encouraged to develop positive thought processes for helping them to cope with pain. They can be helped to recognize negative thoughts about their pain and life activities. Negative thoughts then act as a signal that a positive attempt to change thought processes is required.

Patients must learn to prepare for periods of pain and stress in advance, and develop a coping plan. Positive self statements may help, such as, 'Worrying won't help, what are some of the things that I can do instead? I'm feeling anxious, that's natural, but it's no reason to give up. Just breath deeply and relax.'

When intense stimulation occurs, it has to be confronted and handled effectively. Patients can switch between coping strategies as necessary. Self statements can be used to redirect thoughts to different strategies.

'All right, I'm feeling tense. That lets me know that I should take slow deep breaths and relax, so that I can change to another strategy. I will not let the pain overwhelm me. I will let my thoughts handle the situation just one stage at a time. I will do something positive to confront these sensations.'

If the stimulation becomes intense, the patient must stop the negative feelings of distress and relax, conjuring up a pleasant mental image whilst concentrating on relaxation and breathing. The thoughts can then be turned to concentrating on another mental activity, such as remembering the details of a recent good experience, or a mental exercise requiring great concentration.

Encourage patients to be realistic. They expect periods of intense stimulation, but they should also expect not to magnify them and to keep them under control by using their coping strategies, even though they cannot make the pain go away altogether.

If the patient feels that unpleasant thoughts and negative feelings are beginning to overwhelm them, they must mentally instruct themselves, 'Stop'. It may even help for them to shout this out loudly. The following statements are examples of the self instructions that should follow.

'Things are getting bad. I cannot take any more—no, wait—I should not make things worse. I will review my planned strategies and see what I can switch to. Relax. I will focus my attention on something else. That's better, I am regaining control. A slow deep breath—good.'

'I cannot get my mind off this pain. NO! Wait. I have planned for this. Stop the negative thoughts. Relax, breathe deeply, concentrate the thoughts on one of the coping strategies that I have planned.'

The main point is that the patient learns to recognize negative thoughts, and to interrupt them with positive self statements, so that mental imagery, distraction, and relaxation techniques can be used to cope with the pain. Obviously such techniques need to be tailored for the individual patient and unfortunately, although these methods can be most helpful to many chronic pain sufferers, some patients will be incapable (or unwilling) of developing the necessary skills involved.

References

Elton, D., Stanley, G., and Burrows, G. (1983). *Psychological control of pain*. Grune & Stratton, Sydney.

Fernandez, E. A. (1986). A classification system of cognitive coping strategies for pain. *Pain*, **26**, 141–51.

Fordyce, W. (1973). An operant conditioning method for managing chronic pain. *Postgraduate Medicine*, **53**, 123–8.

Hilgard, E. R. and Hilgard, J. R. (1975). *Hypnosis in the relief of pain*. William Kaufman Inc., Los Altos, California.

Orne, M. T. and Dinges, D. F. (1984). Hypnosis. In *Textbook of pain* (ed. P. D. Wall and R. Melzack), pp. 806–16. Churchill Livingstone, Edinburgh.

Scott, D. L. (1974). *Modern hospital hypnosis*. Lloyd–Luke, London.

Turner, J. A. and Chapman, C. R. (1982*a*). Psychological interventions for chronic pain: A critical review. I. Relaxation training and biofeedback. *Pain*, **12**, 1–21.

Turner, J. A. and Chapman, C. R. (1982*b*). Psychological interventions for chronic pain: A critical review. II. Operant conditioning, hypnosis and cognitive behavioural therapy. *Pain*, **12**, 24–46.

Waddell, G., Bircher, M., Finlayson, D., and Main, C. J. (1984). Symptoms and signs: physical disease or illness behaviour? *British Medical Journal*, **289**, 739–41.

11 Protocols for pain management

These are brief summaries of the systematic management of the conditions most commonly seen in Pain Clinics. Their function is to serve as a check list and not to be a comprehensive guide.

Back pain

Referral

From other medical practitioners, not from self-referral or from other disciplines.

Exclude

Recent onset, rapidly-progressing pain with neurological deficit. Patients with perineal numbness, loss of bowel or bladder control, or recent loss of reflexes need urgent referral to an orthopaedic surgeon or neurosurgeon.

History

Description of symptoms is often detailed but unless it describes objective signs of nerve root compression, it is of little value in pain management. Painful stiffness that resolves during the day may indicate an arthritic element. If present in a young patient, it can indicate ankylosing spondylitis. In these cases the patient should be referred to a rheumatologist.

Examination

Look for scoliosis, loss of mobility, and wasting. Pain on extension rather than flexion suggests arthritis. Vertebral spines tender to pressure may indicate the area from which the pain is arising. Tender paravertebral muscles with localized tender areas may be due to fibromyalgia if the pain is part of a generalized pain problem, or myofascial pain syndrome if the pain is associated with a twitch response to pressure and there are tender bands in muscle. Thorough neurological examination is essential and usually produces negative results. If a previously present reflex has been lost, this indicates that progressive nerve damage is occurring.

Management

Fibromyalgia

Try amitriptyline and exercise, possibly with TENS.

Myofascial pain syndrome

Try trigger spot injections or acupuncture (frequently repeated), cold spray or cold pack with stretching of the numbed muscle, or manipulation techniques.

Other low back pains

If epidural injections have been used successfully before, repeat the previous technique. If pain is obviously in a root distribution give a lumbar epidural, otherwise give a caudal epidural. Give a course of three injections, then follow up at 4 weeks.

If relief is incomplete or absent, look for tender vertebral spines, painful stiffness in the mornings, and sclerotic, narrowed facet joints. Consider facet joint injection, following up at 6 weeks. If there is repeatable but short-term relief from injection, consider radio-frequency facet joint denervation.

If injection techniques are helpful, repeat at the longest feasible intervals. If there is no progress, try stimulation. If pain is localized, low frequency, high intensity electrical stimulation or acupuncture may help. If pain is widespread, try TENS.

If none of these are of help, but the patient is leading an unaltered working and social life, cognitive and relaxation methods may be successful. If the pain is leading to abnormal illness behaviour, with loss of work and restricted social activity, try a pain management programme.

Body wall pain

Referral

Usually from a specialist. Exclude underinvestigated patients and those with symptoms of visceral disease, weight loss, change in bowel habit, exercise induced chest pain, or urinary symptoms. If there is no confirmation of visceral disease on further assessment, these patients should return to the Pain Clinic.

History

Where there is generalized pain associated with tiredness and an altered sleep pattern with deep but unrefreshing sleep, consider fibromyalgia. If there is difficulty in getting to sleep, loss of enjoyment in life, and desperation and tears, consider depression.

Examination

Localized tenderness, with trigger points that produce twitch on pressure and tight muscle bands, may indicate myofascial pain syndrome. Localized pain related to position, exquisite tenderness over the site of emergence of a peripheral nerve from a fascial tunnel, and sensory changes in the distribution of the nerve, may indicate nerve entrapment.

Management

Fibromyalgia

Try amitryptiline, exercise, TENS.

Myofascial pain syndrome

Frequently repeated trigger point injections, acupuncture, or cold spray or cold pack with stretching of tight muscle bands may provide relief.

Body wall nerve entrapment

Confirm diagnosis by injecting a small volume of local anaesthetic, with or without steroid, through an intradermal wheal onto the point of maximum tenderness. Use 1 ml 6 per cent aqueous phenol, repeated at 6-weekly intervals until pain is relieved.

Head pain

Referral

By any medical practitioner.

Exclude

Signs of neurological damage, in particular, recent and progressive eyesight change, signs of CNS infection, haemorrhage, papilloedema, or recent hearing loss.

History

One-sided headache, visual changes with headache, or a throbbing headache associated with nausea indicate migraine or other vascular headache. This can also present as atypical facial or dental pain. Tension headache presents unilaterally or bilaterally, radiating from the occipital to the frontal zone. It was initially stress-related but then became chronic. Jaw pain, associated with one (or all) of back pain, dysmenorrhoea, and an irritable bowel indicate temporamandibular joint dysfunction. Extremely severe lancinating pain triggered by light stimuli and in the distribution of one of the branches of the trigeminal nerve or the glossopharyngeal nerve indicate trigeminal or glossopharyngeal neuralgia.

Management

Vascular headache

Migraine prophylaxis is worth trying in any facial pain of a throbbing nature. If this is unsuccessful, try stress management, trigger spot injections, or acupuncture.

Tension headache

Regular analgesia at the maximum dose should be tried. If this is unsuccessful, trigger spot injection, acupuncture to the tender muscle, and stress management may provide relief.

Trigeminal and glossopharyngeal neuralgia

Try anticonvulsants, changing from carbamazepine to sodium valproate, and reinforcing with carbamazepine. Nerve blockade progressing proximally towards the trigeminal ganglion, trigeminal injection of glycerine, radio-frequency lesion, or, if the patient is fit enough, microvascular decompression have all been used successfully.

Pain in malignant disease

Perhaps more than with any other type of chronic pain, the management of pain in malignant disease depends upon the initial assessment of the cause and nature of the pain. This reflects the fact that there is generally a greater understanding of the aetiology of painful symptoms in malignant disease. It is necessary to ask several questions in a logical sequence in order to plan the best system for dealing with pain in malignant disease.

Underlying pathology

What is the underlying pathology, and why is it producing the symptom of pain? This question leads to a classification of the major types of pain, depending on aetiology.

Nociceptive pain

This can be due to infection, local inflammation, organ or tissue invasion, compression, or distension, or bone erosion or fracture.

Neurogenic pain

This pain may be caused by nerve compression, nerve invasion, or CNS lesion, or be iatogenic (pain which is post-irradiation, post-surgical, or follows nerve block).

Other

Other types of pain include pain produced by coincidental disease, psychological pain or distress, or discomfort from other symptoms.

Management of nociceptive pain

If pain is considered to be at least partially nociceptive, it may respond to analgesics, unless it is mainly incident pain (i.e., pain on movement). If there is an obvious infective factor, then treatment with antibiotics (local or systemic) may be indicated. Wound toilet may be required.

Simple analgesics

For mild pain, try a simple analgesic, such as paracetamol. If there may be an inflammatory component, then the addition of a NSAID may be useful.

Opioid therapy

If a simple analgesic is not effective, or if the pain is already severe, start opioid therapy, with or without the addition of a NSAID. The routes of administration, in descending order of preference, are:

(1) oral/sublingual;

(2) rectal;

(3) subcutaneous infusion;

(4) epidural or spinal;

(5) intramuscular.

Ideally, choose a long-acting opioid, with minimal side-effects, and no ceiling effect. Pure agonists are more versatile.

Increase the dose until it produces effective analgesia, or the drug is no longer tolerated. If there is no evidence of a reduction in pain as the dose is increased, consider:

(1) The pain is opioid insensitive (e.g., incident or neurogenic pain);

(2) The pain is only partially opioid sensitive, and a coanalgesic, such as NSAID, tricyclic antidepressant, or steroid should be added;

(3) Other factors, such as other symptoms, mental pain, and so on.

If opioid is not tolerated, try increasing the dose more gradually, changing to a different opioid, or prescribing other drugs such as anti-emetics simultaneously. Consider changing the route of administration.

Nerve block and other treatments

If pain is localized and amenable to a nerve block without major complications (such as motor block or incontinence), consider nerve block with local anaesthetic (with or without steroid), or a destructive lesion. If nociception arises in the abdominal viscera, consider a coeliac or lumbar sympathetic block.

If pain arises from bone erosion or fracture, consider surgical fixation, radiotherapy, hormonal treatment, nerve blocks, or NSAIDs.

If pain results from visceral compression or distension, consider steroids (brain, liver), autonomic blockade, or antispasmodic drugs.

Neurogenic pain is often accompanied by dysfunction of nerve activity, such as changes in sensation or motor function.

If there is evidence of nerve compression, consider treating the pain with steroids.

Where symptoms include burning, dysaesthesia, paraesthesia, or allodynia, consider tricyclics or sympathetic block.

For shooting or lancinating pain, consider using anticonvulsants.

Repeated nerve block with local anaesthetic and/or steroid may also be beneficial for neurogenic pain.

Management of other symptoms

Consider the management of other symptoms, especially:

(1) nausea and vomiting;

(2) constipation;

(3) insomnia;

(4) pruritis;

(5) dyspnoea and cough;

(6) foul-smelling discharges;

(7) fear, loneliness, despair, anger, depression, and anxiety.

Neurogenic pain

The pathogenesis of a pain characterized as being of neurogenic origin may be obscure. A 'diagnosis' is helpful, but management often has to proceed along empirical lines, with treatments determined by the nature of the symptoms.

Neurogenic, or neuropathic, pains could be considered in several categorizing. Idiopathic or psychogenic pains may mimic neuropathic pains, but are not categorized as such.

Central pain

For treatment of pain with a cerebral or spinal pathology (cerebrovascular accident, tauma, or tumour), consider antidepressants and/or phenothiazine. If these are of no help, add an anticonvulsant or, if ineffective, try opioid antagonist. Also consider TENS or acupuncture.

Degenerative pain

This type of pain, caused by, for example, demyelinating disease or syringomyelia, may be relieved by tricyclic antidepressants and anticonvulsants. Phenothiazine may also help. In peripheral neuropathy, try tricyclic combined with phenothiazine, and consider TENS.

Pain of an infective pathology

Diseases such as tabes and herpes zoster can cause lasting damage. Postherpetic neuralgia can benefit from prophylaxis with nerve blocks, sympathetic blocks, steroids, or antiviral preparations in the acute phase.

In established cases, repeated nerve blocks may help. Prescribe trycyclics and anticonvulsants. If there is a pronounced continual background aching pain there may occasionally be benefit from the use of an analgesic combined with antidepressants. For dysaesthesia and allodynia, consider topical local anaesthetic or anti-inflammatory preparations. TENS may be of occasional value.

Traumatic pain

Pain from scars, 'trapped nerves', or neuromata may be treated by repeated infiltration of local anaesthetic (LA or LA and steroid), neurolytic infiltration of the scar, trigger point, or neuroma, cryotherapy or radio-frequency lesion to the neuroma, tricyclics, TENS, regional sympathetic block, or physical therapy.

Deafferentation pain

This includes plexus avulsion, phantom pain, and pain from post-neural block. Treatment is by tricyclic and anticonvulsant, sympathetic block, TENS or acupuncture, hypnosis, or by neurosurgical procedures.

Autonomic dysfunction

This category includes autonomic dystrophy, causalgia, algodystrophy. Treat by autonomic block (LA to ganglia, intravenous guanethidine, systemic drugs), ketanserin, acupuncture, or exercise. Tricyclic antidepressant may relieve dysaesthesia.

Trigeminal neuralgia may be relieved by anticonvulsants (in sequence and combination), nerve block with LA, Gasserian ganglion block with alcohol, glycerol, radio-frequency lesion, or microvascular decompression.

Index